Aeration, Mixing, and Energy: Bubbles and Sparks

Aeration, Mixing, and Energy: Bubbles and Sparks

Edited by
Diego Rosso

IWA PUBLISHING

Published by **IWA Publishing**
Alliance House
12 Caxton Street
London SW1H 0QS, UK
Telephone: +44 (0)20 7654 5500
Fax: +44 (0)20 7654 5555
Email: publications@iwap.co.uk
Web: www.iwapublishing.com

First published 2018

British Library Cataloguing in Publication Data
A CIP catalogue record for this book is available from the British Library

ISBN: 9781780407838 (print)
ISBN: 9781780407845 (eBook)
ISBN: 9781789060195 (ePUB)

Contents

About the Editor

Diego Rosso is Professor of Civil and Environmental Engineering and of Chemical Engineering and Material Science at the University of California, Irvine, where he is also Director of the Water-Energy Nexus Center. Since 2000, he has been investigating aeration systems and the water-energy-carbon nexus of water reclamation and reuse processes. He holds degrees from the University of Padua and from the University of California, Los Angeles.

List of Contributors

Linda Åmand
IVL Swedish Environmental Research Institute
linda.amand@ivl.se

Manel Garrido-Baserba
Lecturer and Postdoctoral Scholar
University of California Irvine
manelg@uci.edu

Ken Brischke
Process Engineer Wastewater Technical Leader
AECOM, Inc.
brischkek@aecom.com

Bengt Carlsson
Uppsala University
bengt.carlsson@it.uu.se

José Jimenez
Vice-President
Brown and Caldwell
JJimenez@brwncald.com

Daniel Nolasco
President
Nolasco y Asociados, S.A.
daniel@nolasco.ca

Gustaf Olsson
Professor Emeritus
Lund University
gustaf.olsson@iea.lth.se

Coenraad Pretorius
CDM Smith
pretoriusc@cdmsmith.com

Leiv Rieger
President & CEO
inCTRL Solutions Inc.
rieger@inctrl.ca

Michael K. Stenstrom
Distinguished Professor, UCLA
stenstro@seas.ucla.edu

Gan Wang
President & CEO
PLAKeco, Ltd.
wanggan@vip.163.com

Preface

This book presents the state of the art on aeration technology for water and wastewater treatment systems. Aeration, or oxygen transfer, is the heart of wastewater treatment and must be performed well and at high efficiency for successful treatment.

The very first attempts at aerobic treatment by Ardern and Lockett showed the need for aeration, even though they did not fully understand the theory of biological oxidation. As more knowledge developed, aeration was always provided and it was soon understood that the oxic conditions provided by aeration were essential in any type of wastewater stabilization.

The 1950s saw a rapid development in understanding of aeration and advancing aeration technology. Large cities with concentrated wastewater discharges faced the growing need for sewage treatment, including secondary treatment using the activated sludge process. Large activated sludge processes were constructed in cities that did not have the benefit of coastal outfalls for dilution, which was at the time the preferred mode of sewage disposal. These installations soon found they needed not only practical aeration technology, but also better methods to quantify aeration requirements as well as specification of aeration equipment. An early summary of aeration can be found in the proceedings of the 1st Manhattan College Conference on Biological Treatment of Sewage and Industrial Wastes, held in 1955. Part 2 of that volume provided a basis for aeration theory and design, written by pioneers in our field, and much of this early analysis is still current. The 3rd Manhattan College Conference reported on the development of new research to understand and quantify the effects of soaps and detergents on aeration systems, as well as descriptions of new aeration equipment.

From that time on, it was common to find coarse-bubble diffusers in 'spiral roll' configurations at large treatment plants. These systems provided excellent mixing but are now recognized as the least efficient method of aeration for the activated sludge process. Development continued with slightly modified geometries, described as 'cross-roll' or 'ridge and furrow'. The development of more efficient systems was mandated by the rapid increase in energy costs and development of fine-pore aeration systems, with early development in Britain, later arriving elsewhere and now becoming commonplace in municipal aeration systems. It is practical now to use third-generation fine-pore aeration systems that are four times as efficient as early spiral roll systems.

Today we find the most common aeration systems to use 'full floor' coverage of tank bottoms with floor coverage as large as possible. Deeper aeration tanks can now benefit from more efficient blower technologies. Aeration systems are now designed not only for improved efficiency but to reduce stripping of volatile organic compounds as well as controlling heat loss.

Decades after that 1st Manhattan College Conference, we recently presented there our work that in part is summarised in this book. After reviewing the most current aeration systems, this book presents the best ways of measuring aeration system performance and to use those measurements for design, control, and sustainable operations. A team of experts in the field were assembled to help write this book, which is the product of several years of work and of the immense patience of our publishers: in fact, it took us long enough to finish this book that we managed to retire one of them, almost two.

By no means do we intend this to be the ultimate encyclopaedia on bubbles and sparks. We recognize that, in order to complete the book and avoid seeing the retirement of the second publisher, we had to draw a line and submit our manuscript for publication.

Diego Rosso
University of California, Irvine

Michael K. Stenstrom
University of California, Los Angeles

Chapter 1

Aeration equipment

Ken Brischke[1], Diego Rosso[2] and Michael K. Stenstrom[3]

[1]*AECOM*
[2]*University of California, Irvine*
[3]*University of California, Los Angeles*

1.1 INTRODUCTION

In this chapter we present a *very brief* survey of the main equipment components of an aeration system. First of all, let us clarify that none of us sell equipment. Equipment manufacturers and vendors should be the reference point to describe their own technology. What you see here is a presentation of our observations in our experience. We do not intend to take sides on any of the technologies discussed here but rather present a report on the phenomena affecting their operation and performance.

1.2 THE BLOWER HOUSE

Blowers are mechanical devices used to produce relatively high volumes of low-pressure air or gas for mixing, aeration or combustion purposes. These adiabatic machines impart velocity to the airflow, just like a hairdryer. However, we need to clarify at the outset a critical point that will be associated with aeration systems throughout the book:

Blowers do not compress air, they blow it.

Editor: Diego Rosso
doi: 10.2166/9781780407845_0001

Why is this concept so important? Because we will see later how the limitations in the blower discharge pressure can be a crippling constraint in the selection and operation/maintenance of diffusers. The adiabatic compression operated by blowers has a corresponding increase in pressure at the outlet of the blower only in virtue of the fact that we impose a pressure drop onto the aeration system. *No pressure drop, no compression* (in fact, the pressure at the outlet of a hairdryer is very similar to the inlet, since there is no pressure drop imposed: only the velocity of the air leaving is very different than the air in the surroundings of the outlet).

And what does *adiabatic* mean? It means that the compression is performed without heat exchange. For an ideal blower this means that the air in the discharge piping (as hot as it may be due to its compression) will return to the same temperature of the inlet once it is released to its initial pressure (i.e., when the air transits through the aeration diffusers, as we will describe later in the book). In reality, some heat losses from a hot steel pipe to the surrounding atmosphere and in water should be expected, and therefore we often see the air distribution pipes jacketed. Were they not, in a cold winter the heat loss would be such that the air bubbles would be released at temperatures much lower than the surrounding (freezing) atmospheres, thereby posing potential thermal problems within the reactors.

Blowers must satisfy the aeration and mixing requirements at various locations within a water resource recovery facility (WRRF), such as aerated grit removal systems, fixed-film/suspended-growth systems, activated sludge processes including membrane bioreactor processes, aerobic digestion, and post aeration systems. Blowers provide low-pressure air for process needs. Why low pressure? Because of the blower rule we wrote above. Since their ability to overcome a pressure drop is very limited (remember, they are not compressors!), at some point every blower will meet its maximum discharge pressure, which is referred to as *surge*. The surge zone is always on the upper left area of a blower curve (see below). When a blower surges, it can no longer displace air through the outlet and begins an unstable operation (vibrations begin rapidly, due to the need for the blower to exert somewhere the mechanical work imparted by its motor) until some safety controllers impose a shutdown. A surge event at a WRRF does not go unnoticed: the noise become noticeably loud and the gratings on the blower building begin vibrating (and at times may fall off their anchors). At home you could try running a hairdryer on cold air and place your hand on the discharge outlet: after a while, the motor spinning in vain in stagnant air may burn out (in which case do not ask us for a reimbursement).

The compressed air is distributed through a piping system to submerged diffusers for release at the floor of the reactor. As the bubbles rise through the bulk liquid, oxygen is transferred into it. A well designed blower system supplies a wide range of airflows, matching fluctuating air demand, typically with a relatively narrow pressure range under variable environmental conditions.

The main blower applications can be divided as follows:

Process (secondary) aeration. The biggest user of power, and by extension air, is secondary treatment. Aeration is used to transfer oxygen into the bulk liquid for use by the active biomass during the degradation of organics and the oxidation of ammonia. At the same time, mixing energy associated with aeration is used to keep biomass and media, if present, in suspension.

Aerobic digesters. Similarly to aeration in secondary reactors, blowers provide air for oxygen mass transfer and mixing requirements. As sludge is added to and withdrawn from aerobic digesters, the liquid level in the digester can vary, resulting in variable discharge pressure requirements and airflow requirements.

Anaerobic digesters. Gas mixing in anaerobic digesters typically uses a constant gas flow.

Mixing. Channel aeration (mixing), aerated grit basins and other uses. Air could also be used to mix unaerated zones in the activated sludge reactor.

As power costs continue to escalate in many areas worldwide, with aeration being the major on-site electrical consumer, there is an ongoing search for more efficient aeration systems. Blowers, as key parts of aeration systems, need to be carefully evaluated. Particular attention must be paid to their 'turn-up' or 'turn-down' capabilities to ensure that the full range of demands can be met, without any unattainable flows in that range. Remember that 4/5 of the air is composed of a gas that is not contributing to aeration, hence even more attention must be driven to the transfer of oxygen: any changes in oxygen requirements multiply several-fold as air requirements.

1.2.1 Types of blowers

Blowers are classified into two broad types: positive displacement and dynamic (typically referred to as centrifugal). There are a number of different configurations of each type of blower as shown in Figure 1.1.

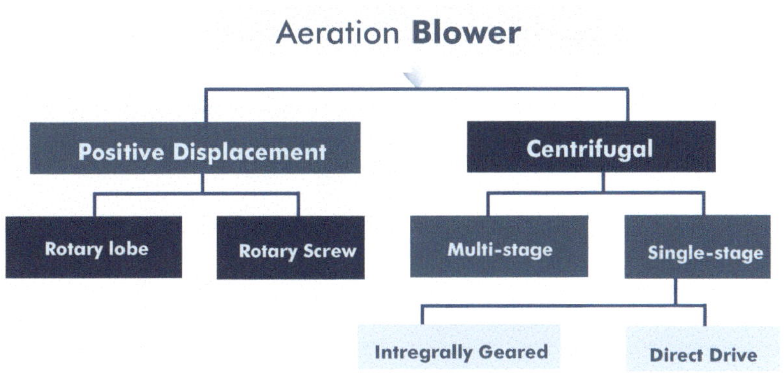

Figure 1.1 Positive displacement and dynamic blower configurations.

Positive displacement (PD) blowers are generally considered as constant-flow and variable-pressure devices. PD blowers continuously compress a fixed volume of gas in an enclosed space to increase the pressure and provide some velocity. On the other hand, centrifugal (or dynamic) blowers, like centrifugal pumps, produce variable flow and pressure, using rotating impellers. There are advantages and disadvantages of the two types, which are summarized in Table 1.1.

Table 1.1 Summary of features characteristic of PD and centrifugal blowers.

Positive Displacement	Centrifugal
• More economical at small scale	• Economical at all scale but especially for large installations
• Noisy – the rattling associated with the rotary lobes is harder to dampen. Three-lobe blowers partially address this shortcoming	• Can also be noisy but the continuous, higher-frequency spinning sounds are easier to dampen
• Vibration transmission to piping and supports sometimes problematic • Motor overloads with excessive discharge pressure, requiring current protection on motors	• Operation at excessive flow overloads the motor, and operation at excessive pressure causes surge, which may result in destruction of the blower. Current-protection and vibration-detection controls are required for safe operation
• Higher discharge pressures possible	• Higher discharge flow possible

At smaller plants, PD blowers are often an attractive option, while centrifugal blowers are mostly used for secondary treatment aeration at larger plants. Before the advent of efficient variable-frequency drives (VFDs), there was little opportunity to modulate the flow of a PD blower. Some energy could be consumed by throttling the suction or, in other cases, excessive discharge flow could be vented at reduced pressure. Neither situation was very satisfactory. With VFDs, the flow is proportional to blower rpm (less a small fraction due to slippage), and a wide range of turn-up or turn-down is possible.

1.2.1.1 Positive displacement blowers

Positive displacement blowers produce compressed air at constant capacity and varying pressure. PD blowers compress discrete 'packets' of air with a couple of bi- or tri-lobed gears or rotors. The following drawing illustrates the concept (Figure 1.2). Due to the discrete nature of the process, positive displacement is not efficient as centrifugal flow, but can achieve higher output pressures for the same airflow rates. Also, the airflow can be varied by varying the speed of the PD blower. The disadvantage of this type of blower is the noise produced by compression, typically recognizable as a loud rattling or tapping noise.

Figure 1.2 Photograph of a positive displacement blower.

As mentioned, PD blowers are not perfect, and losses reducing the efficiency occur inevitably. Losses in positive displacement blowers occur as air under pressure escapes around the lobes, and between the casing and the lobes.

Rotary lobe

The typical rotary lobe blower, as shown in Figures 1.3 and 1.4, consists of two symmetrical impellers that rotate in opposite directions in an enclosed space,

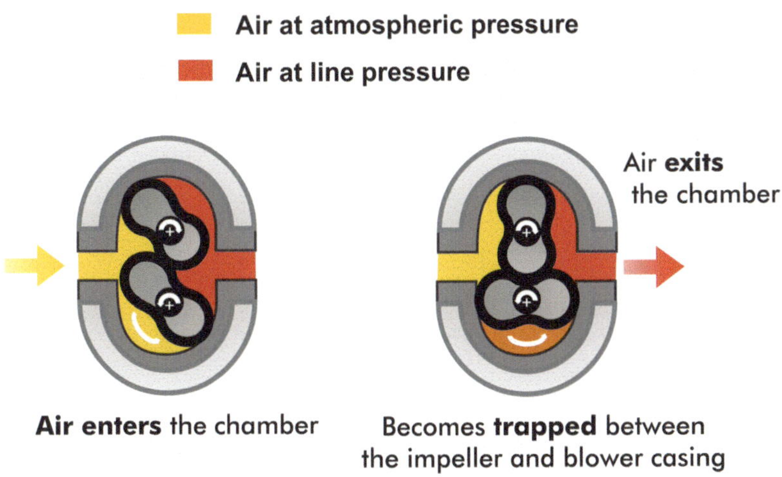

Figure 1.3 Schematic of a positive displacement blower.

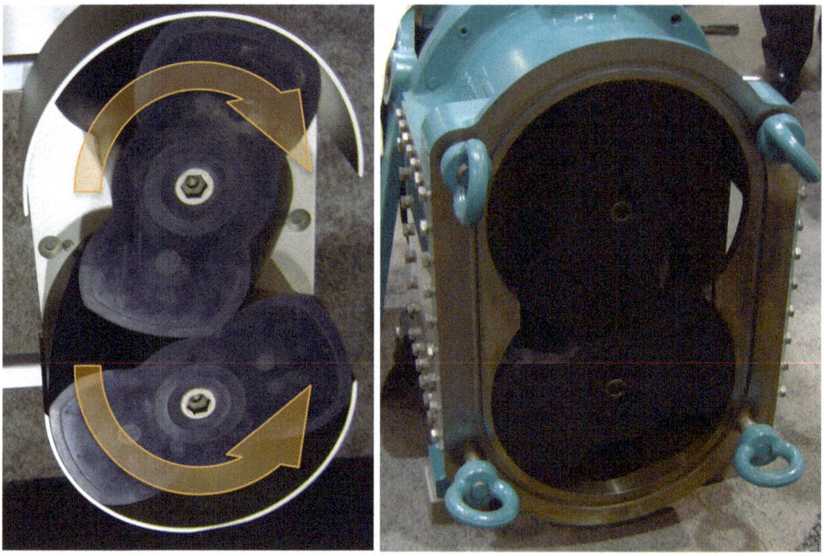

Figure 1.4 Rotary lobe positive displacement blower.

moving gas from the inlet to the outlet. Rotation of the lobes does not compress the gas. Pressure is generated by restricting discharge of the gas from the machine.

Rotary screw

Similar to the rotary lobe blower, the rotary screw consists of two helically shaped shafts rotating in different directions in an enclosed space. The cavity volume between the two shafts progressively decreases resulting in an increase in pressure and decrease in discharge capacity from the inlet to the outlet of the blower casing. Unlike in the traditional PD blower, the rotary screws operate in continuous mode (not discretely) and are *de facto* a hybrid technology between PD and centrifugal.

1.2.1.2 Centrifugal blowers

Centrifugal blowers are more common than positive displacement blowers at larger plants due to larger volumes of low-pressure air available.

Centrifugal blowers rely on the rotational speed of an impeller within an enclosed casing to provide velocity to the gas. The air is continuously discharged radially and its increased kinetic energy is converted to a pressure increase by imposing the pressure drop on the discharge. Figures 1.5 and 1.6 illustrate the concept. Also, original centrifugal blowers do not have turn-up or turn-down capability, and have to be operated at constant rotational speed.

Newer technologies include centrifugal blowers with variable intake guide vanes and variable outlet vanes. By varying the angle of the vanes, the airflow rate can be

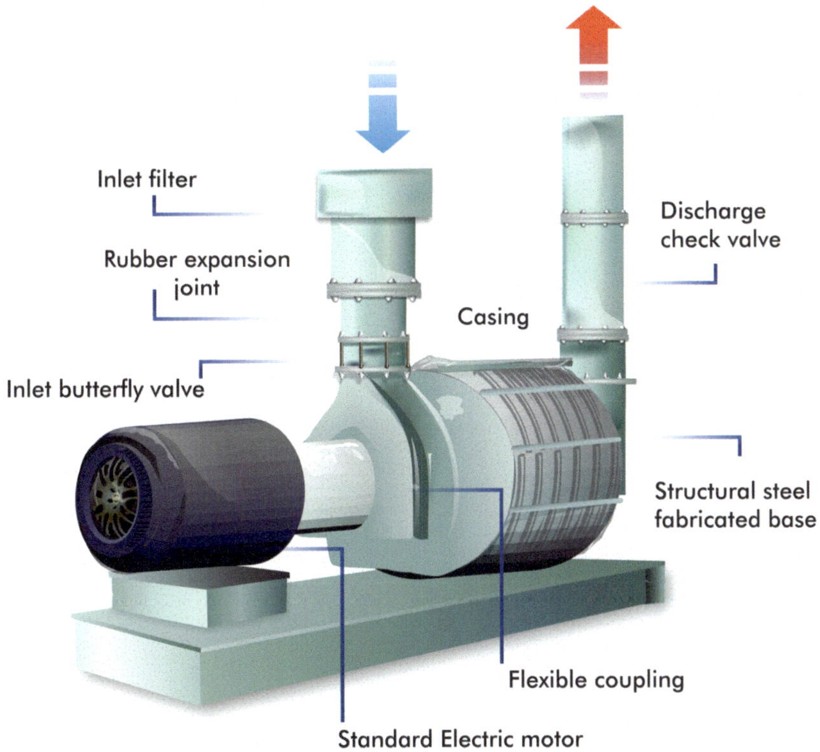

Figure 1.5 One of the most common blowers in WRRFs: A Multistage Centrifugal Blower.

varied and the blower acquires turning capability. Nevertheless, centrifugal blowers have an optimum operating region, and outside that region efficiency declines.

As with positive displacement blowers, centrifugal blowers are also subject to losses as gas under pressure escapes between the impellers and casing and around the impellers. Figure 1.7 shows the discharge flow and pressure relationship of a number of commonly used blowers.

Blower performance is affected by changes in the temperature of the inlet gas and barometric pressure which cause a change in density of the compressed gas. The density of the gas or air increases with lower temperature and/or higher pressure, requiring greater power input to compress the same volume of gas.

Multi-stage centrifugal blowers

A typical multi-stage centrifugal blower is shown in Figure 1.8. Under conventional operation, multi-stage blowers use a number of impellers in series to increase the discharge pressure of the gas. In this configuration, variable output can be

Figure 1.6 Image of a single-stage centrifugal blower.

achieved by installing a control valve on the blower inlet or through the use of a variable-speed drive on the blower motor. Turn-down of multi-stage centrifugal blowers is limited however because their curve is flat as output approaches the surge point as shown in Figure 1.9. Figure 1.10 shows typical performance curves for multi-stage centrifugal blowers with a variable-speed driver. Note the

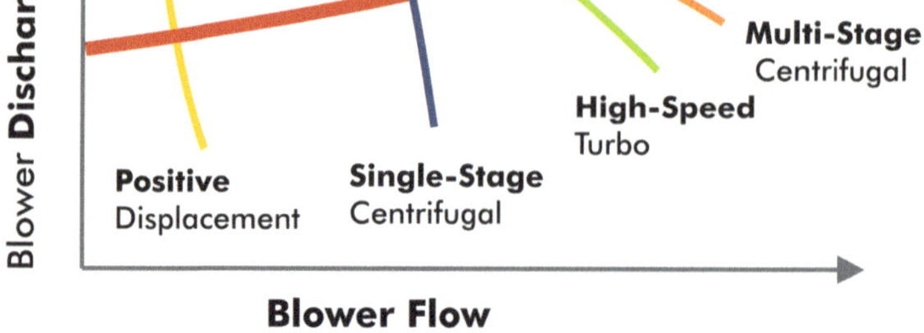

Figure 1.7 Pressure vs Flow relationship for Commonly Used Blowers (adapted from Loera, 2013).

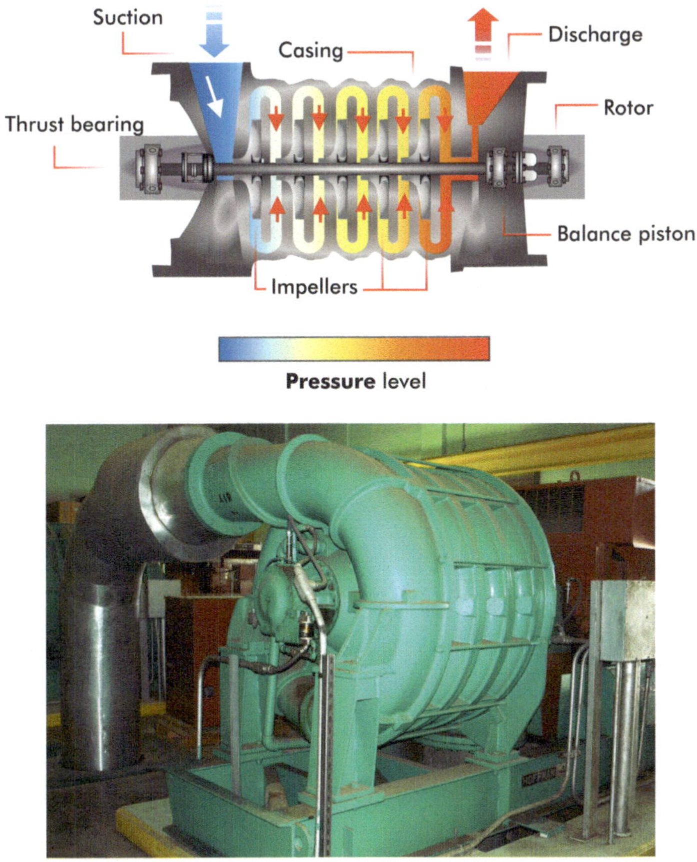

Figure 1.8 Multi-Stage Centrifugal Blower schematic (top) and Example of a multi-stage centrifugal blower in operation at a large wastewater treatment plant. The air discharges from the top of the blower casing, as indicated by the label on the discharge pipe.

proximity of the anti-surge control line to the predicted surge limit. The performance curves become quite flat as the capacity is reduced towards the surge line, creating a zone of potential instability. It is important to ensure that the anti-surge control line is located sufficiently far from the predicted surge line to maintain stable operation. Location of the anti-surge line identifies the minimum turn-down of the blower.

Single-Stage Centrifugal Blowers

Single-stage centrifugal blowers rely on a single impeller, as shown in Figure 1.8, to develop the required discharge pressure and airflow volume. There are a number of different single-stage centrifugal blower configurations (as shown in Figure 1.1).

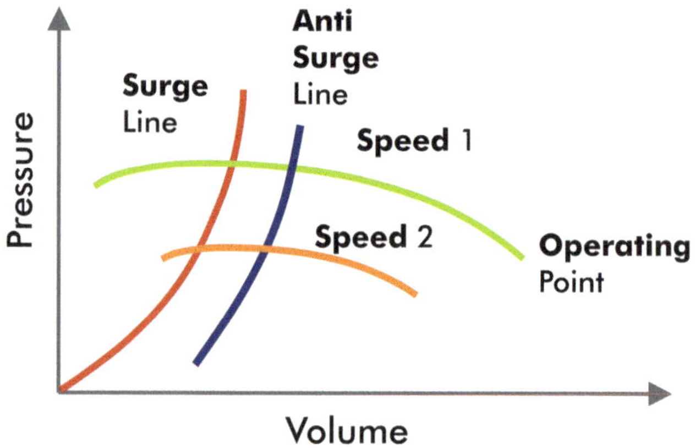

Figure 1.9 Multi-stage centrifugal blower performance limits.

Differences in the various configurations include variable-frequency driven (VFD) motors versus constant-speed motors, or the use of inlet guide vanes and outlet diffuser vanes.

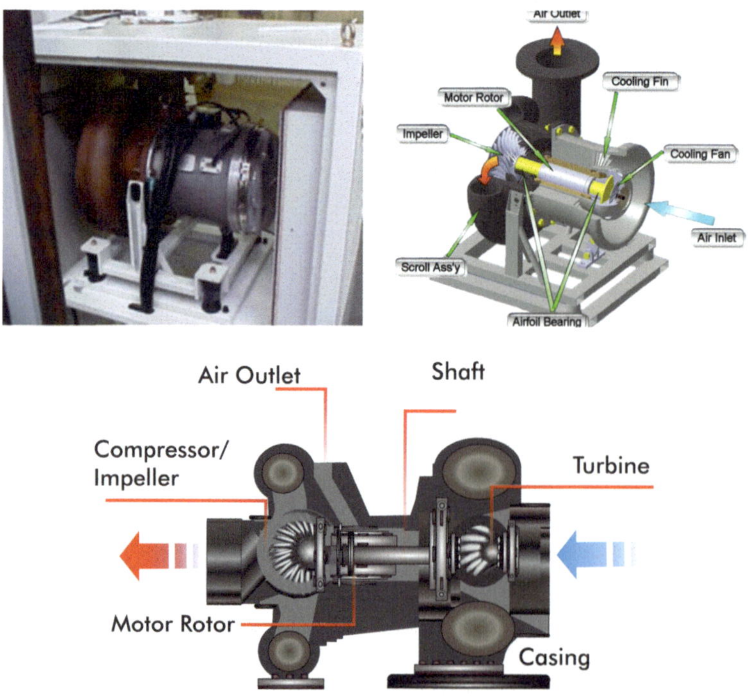

Figure 1.10 Direct-drive or high speed turbo blower.

Direct-drive or high-speed rurbo. The direct-drive, or high-speed turbo single-stage centrifugal blower utilizes an impeller very similar to the integrally geared single-stage blower impeller. They are referred to as direct-drive because the impeller is attached directly to the motor armature shaft. They are also referred to as 'high-speed turbo' because their rotational speed is very high (as high as 70,000 rpm). Because of the high rotational speed, specialized equipment such as a high-speed VFD, an EMC filter, a harmonic filter, and a sinusoidal filter, etc., becomes required to complete the full blower package. These blowers were introduced to the market in the early 2000s. Direct-drive single-stage centrifugal blowers are manufactured in two configurations. Owing to the way the motor and impeller are closely integrated, direct-drive blowers are ideally suited to 'plug-and-play' installations.

Figure 1.10 shows the horizontal configuration using airfoil bearings. Since it uses airflow to suspend the rotor axis, the airfoil bearing type is not a good choice for applications requiring frequent stops and starts. In fact, this is the bearing for the turbines of jet airplanes and that is why aircraft in airports do not use turbine power (to avoid wear of the shaft). The other option is to use permanent magnet bearings. The capacity of these blowers is currently limited by the bearing technology and the weight of the armature.

Integrally geared. An integrally geared single-stage blower relies on a gearbox located between the drive motor and the blower to increase the speed of the specially designed impeller to provide the required airflow and discharge pressure. The gear box and impeller are designed based on variable site-specific conditions of airflow, discharge pressure, inlet air temperature, and relative humidity. Figure 1.11 shows a cut-away view of an integrally geared blower showing the airflow path through the inlet guide vanes, into the impeller, through the outlet diffuser vanes to the discharge pipe. The blower operates at fixed rotational speed and the complex control system that continuously varies the angle of the inlet and guide vanes allows for variable flow and relatively high efficiency even at turn-down.

1.2.2 Airflow control

Two key parts on a typical blower can be used to control the rate at which the air is delivered.

Variable-frequency drives (VFD). A variable-frequency drive (VFD) is an electronic system that allows the control of the frequency for alternating current (AC), therefore controlling the rotational speed of the electric motor connected. As an example, if an electrical motor has a speed of 1500 rpm with an electrical frequency of 50 Hz, its speed can be reduced to 1200 rpm by reducing the frequency to 40 Hz; in other words, the rotational speed of the motor and blower are directly proportional to the frequency of the power supply. Fan laws (also known as affinity laws) suggest that air or gas flow is directly proportional to

Figure 1.11 Integrally geared single-stage centrifugal blowers. Top left: blower battery; top right: detail of intake while on display; center: detail of the servo motor operating the guide vanes; bottom left: detail of the inlet vanes; bottom right: detail of the discharge guide vanes.

rotational speed of the blower, while pressure is proportional to the square and power to the cube of the rotational speed, respectively. Thus, reducing the frequency by 20% would reduce flow by 20%, pressure by 36% and power usage by 49%, so that significant power savings can be achieved with modest reductions in frequency. Note that system pressure is determined by the system curve, not the blower curve. Also note that the VFD itself is not an ideal unit, therefore the energy savings should be lower than the affinity laws project.

By installing a VFD, the motor can be run at higher or lower speed than its nominal rating, and can be started and stopped gradually with less overheating. When conventional motors are started, several times the rated current is initially drawn to bring the motor to speed (referred to as inrush). This generates significant heat and for large motors may limit the ability of the motor to be restarted more than once in an extended period of time (e.g., not more than once per hour, to provide time for cooling). At the same time the high starting current may trigger power demand penalties, especially if started during periods of peak daily power usage. Demand charges and penalties can be a significant part of the power bill. Diurnal cycles in wastewater treatment typically result in highest treatment requirements during daytime or afternoon, when power may have a higher cost. Therefore, reducing the power drawn for motor start-up can result in extended equipment life and significant cost savings.

Inlet and guide vanes. Fixed or variable vanes, also known as stators, are located at the inlet (Figure 1.12). These vanes direct the air onto the rotor at the optimal angle for a given flow. The basic purpose of the inlet vanes is to direct the fluid on to the runner blades at the angle appropriate for optimal operation. The guide vanes, placed on the discharge ring, adjust their angle and provide control on the airflow. By varying the angle of the guide vanes, the airflow rate can be varied and the blower acquires tuning capability.

Figure 1.12 3-D model of a single-stage centrifugal blower with variable guide (grey) and distribution (red) vanes displayed at a trade-show.

1.2.3 Application and selection

In wastewater treatment plant operations, blowers are typically required to provide a wide range of airflow volumes over a narrow pressure range, often under widely varying atmospheric conditions. It is important to ensure the machine is correctly paired with the application.

Selecting the correct blower type for a specific application is the first step for blower selection. Generally, each of the two main types of blowers is more suited to specific processes. While positive displacement blowers are well suited to provide air to processes exhibiting variable liquid levels and less variable airflow requirements (such as sequencing batch reactors and aerobic digesters), centrifugal blowers are well suited for applications requiring more varying volumes of air at relatively constant discharge pressure. However, further guidance for the selection of PD and centrifugal units can be found in Table 1.1.

Furthermore, aeration airflow rates can vary widely throughout the day and from day to day throughout the year. Diurnal airflow rates can vary significantly throughout the day (see Sec. 6.3). With stringent effluent requirements for total nitrogen and phosphorus, strict control of dissolved oxygen concentration becomes more important to minimize dissolved oxygen contamination of anaerobic and anoxic zones, often reducing the airflow rate required for respiration below the traditional mixing requirements (see Sec. 3). This often requires different-size blowers to be installed to meet the minimum and maximum airflow rates, as it is unlikely that a single blower model can match the full variation in airflow rate throughout the year, without installing a prohibitively large number of units. Installing different-sized blowers also reduces the risk of having unattainable flows throughout the required range of flows. The fact that start-up flows and loads may be much lower than design flows and loads, and design flows and loads sometimes never materialize, is another reason to consider multiple blower sizes.

Other factors include life-cycle cost and flexibility of operation. Both of these factors are somewhat interrelated. Life-cycle cost is a combination of capital plus operation and maintenance costs. Figure 1.13a shows a generalized comparison of the various blower types for efficiency versus flexibility. At the current cost of power, equipment cost ranges from 10 to 15% of the life-cycle cost over a 20-year analysis period, which means power costs make up the remaining 85 to 90%. Efficiency of the blower itself thus becomes an important selection factor, and will be more so as power costs increase.

However, efficiency and flexibility are not the only two dimensions one must consider when selecting and operating blowers. In Figure 1.13b we report a different generalized plot where the process requirements (pressure and flow) are on the axes. Clearly, the placement of the blower technologies on this plot is very different. Therefore, multiple criteria must be considered when selecting a blower.

Historically, blowers were specified and selected based on a few performance points such as average day, maximum weekday and maximum day. Dynamic process simulators can now be used to generate hourly airflow rates allowing for a more detailed evaluation process comparing a significantly larger number of

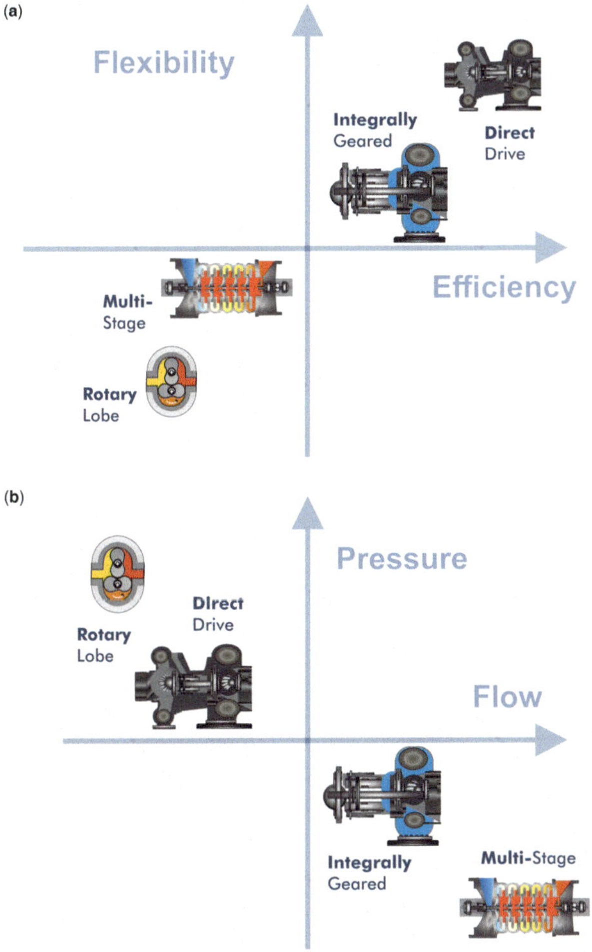

Figure 1.13 Blower type comparisons: (a) flexibility versus efficiency; (b) flow versus pressure.

performance points. Designers can select specific days representing different airflow requirements such as:

- Minimum Day at start-up in winter
- Average Day at both start-up and design conditions
- Maximum Month at design
- Maximum Week at design
- Average Summer Day at design
- Average Winter Day at start-up

In association with the required airflow rate, site elevation, inlet air temperature and relative humidity are the other main factors in determining blower power requirements (see Chapter 6).

1.2.4 Blower testing

Consistent testing is important for fair comparisons. For example, ASME has developed a new test code, PTC 13: Wire-To-Air Performance Testing for Blower Systems, to standardize the testing of *all* blowers. The predecessor to PTC 13, the Test Code for Displacement Compressors, Vacuum Pumps, and Blowers, now designated as PTC 9, was originally issued in 1915. The Test Code for Compressors and Exhausters, PTC 10, was originally issued in 1934, followed by several editions. The main shortcoming of PTC 9 is that it requires measuring the torque transferred by the motor to the blower – since this axis is not accessible in direct-drive blowers, the torque cannot be measured, and the standard cannot be used to verify the efficiency of these blowers.

Elsewhere, analogous testing standards may exist. As for diffusers, it is paramount that blower testing be conducted according to standard protocols and codes, under the witnessing eyes of independent experts.

1.2.5 Blower upgrades and recommendations

When considering blower upgrades, several factors should be taken into consideration. Blowers must be chosen, accounting for redundancy, to allow scheduling shifts and operations and maintenance requirements. In order to avoid sudden increases in airflow rates (and therefore in power demand), blowers with soft starting capability are always recommended, but tuning capability is preferred (i.e., positive displacement blowers with variable-frequency drives, or centrifugal blowers equipped with guide vanes, outlet diffusers and/or variable-frequency drives, etc.). These blower systems allow variation of the airflow within their operating range, which accommodates the variations in demand by the treatment process. When the air demand exceeds the blower capabilities, another blower is activated, as in traditional systems. The benefit of tuning systems is greater flexibility and smoother transition within the range of airflow rates, which facilitates improved process control and minimization of energy costs.

A classic problem that haunts operators and process control engineers is 'hunting', which occurs with dissolved oxygen (DO) control systems. The basic problem is that the blower is treated as a non-limited continuous source of air by the control algorithm. An example explains it best.

A treatment plant is composed of several, parallel aeration tanks. When one tank has low DO, caused by a flow imbalance or random effect, the controller calls for more air and opens an air valve, which provides more air to the affected tank. Ideally, the additional air should be provided by the blower, but in reality it is not. Instead it robs this additional air from the air supply of an adjacent tank. This occurs because of pressure drops in the air distribution system and because of the nature of the blower. The loss of air in the adjacent tank causes the DO to drop, and the controller calls for even more air, which robs air from other tanks. Eventually all tanks are calling for more air and the control system finally responds by turning on an additional blower. Because the blowers have pre-set ranges of flow, and not a continuous distribution of flow rates, the air to all tanks increases and the DO values begin to increase. One tank will be first to reach excessive DO and the control system will reduce the airflow. This does not reduce blower output, but only forces more air into other tanks. Very quickly all the tanks begin to have excessive DO, and the control system finally turns off the additional blower. Now the cycle starts over again and the DO values will decline until the blower is turned back on again, when all tanks will have excessive DO, yet again. This is because the control system keeps 'hunting' for a solution it cannot find.

As we will see in Chapters 4 and 5, the problem can be addressed, in part, by using a cascade control where the DO error signal (master control loop) is used to adjust the airflow set-point and airflow is the slave control loop. This minimizes the impact of distant valve adjustments to the airflow in a given drop leg. Cascade control does introduce some lag between changes in the DO concentration and valve position response. There are commercially available control systems that address these challenges. These systems continuously track oxygen transfer efficiency, so that the exact relationship between the required DO adjustment and the required airflow is known and constantly being reevaluated.

The impact of 'hunting' is excessive energy consumption for starting and stopping of blower motors as well as an increase in wear and tear on the blowers, valves, and actuators. In cases where the operators become concerned about the impacts on plant performance, they may disable the DO control system altogether and cause over- or under-aeration. Usually, operators choose over-aeration to avoid losing nitrification, which may take a long time to recover from; instead they risk violating nutrient limits or using excessive chemical dosing to compensate. Later in the book we will discuss in more details the pitfalls of over-aeration.

In conclusion

- The choice of air blowers and air distribution systems is a substantial capital investment and has consequences on operating costs throughout the lifespan of the wastewater treatment plant.
- However, the *iceberg* rule applies here: the capital cost is the portion above the water and the electricity to operate the unit over its useful life is the portion below the water.
- The design engineer must confirm turn-up turn-down capability of the blowers under all operating conditions, and consider including centrifugal blowers with variable guide vanes or outlet diffusers, positive displacement blowers, and variable-frequency drives.

- DO control systems may fail for a number of reasons, resulting in 'hunting', which is the continuous search for an optimum set-point. Hunting results in increased wear and tear of the aeration and control systems. In many cases, operators set the aeration system at an arbitrary high operating point to bypass hunting, but then facing the pitfalls of over-aeration.

Recommendations

- Care must be used when choosing blowers. Blowers with turn-up turn-down capability should always be evaluated as an alternative. This should be considered for both new designs and retrofits of existing installations.
- Initial savings in capital cost may result amplification in the operating costs and in the latitude of operation. The possible higher capital cost of newer blowers should be compared in a net present worth analysis with the increased operating cost of traditional blowers. Also, in this analysis, different growth scenarios for air demand should be considered, and the limitations of conventional air blowers should be accounted for. In certain areas where power utilities offer rebates for equipment upgrades, blower upgrades should always be considered.
- To mitigate hunting, several changes are needed. The first is to provide blowers with larger turn-up and turn-down capabilities. The second is to provide a 'smart' control system such as continuous monitoring of oxygen transfer efficiency. This requires that the control system be equipped with an accurate estimate of the airflows needed to match the oxygen demand and the air valves on all tanks can be adjusted appropriately.

1.3 THE AERATION TANK

The next section presents an overview of commercially available aeration systems (excluding the blowers), whose characteristics will be described later in the chapter. Figure 1.14 illustrates the main components of the aeration system near or within the tank. Table 1.2 summarizes of the defined terms.

1.3.1 Mechanical aeration

1.3.1.1 Surface aerators

Surface aerators belong to the first generation of oxygen transfer technologies. These aerators may or may not need a blower to perform oxygen transfer. Surface aerators shear the liquid into small droplets which are spread in a turbulent plume at several metres per second. The travelling droplets are in turbulent contact with the atmospheric air and typically oxygenate to at least half-saturation. As soon as they land on the liquid free surface they mix with the liquid bulk, producing a typical DO pattern as in Figure 1.15. Since there is no airflow with the surface aerators, the OTE can neither be defined nor measured. The circulation of the

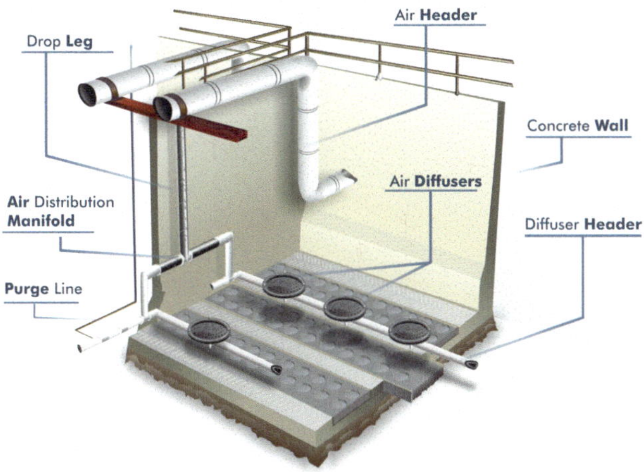

Figure 1.14 Representation of the different parts in aeration systems.

liquid drawn into the spray also provides mixing. In some cases, surface aerators are also specified by the liquid pumping rate as well as an OTR. Surface aerators are typically characterised by high OTR and low SAE values (in the range of 0.9–2.1 kg O_2 kWh^{-1}).

Table 1.2 Summary of the nomenclature used in this book.

Parameter	Definition
OTR	Oxygen transfer rate in clean water
SOTR	Oxygen transfer rate in standard conditions in clean water
OTE	Oxygen transfer efficiency in clean water
SOTE	Oxygen transfer efficiency in standard conditions in clean water
AE	Aeration efficiency in clean water
SAE	Aeration efficiency in standard conditions in clean water
k_La	Liquid-side mass transfer coefficient
α	Alpha factor, that is ratio of process to clean water mass transfer
F	Fouling factor
αF	Alpha factor for used diffusers
αSOTE	Oxygen transfer efficiency in standard conditions in process water
αFSOTE	Oxygen transfer efficiency in standard conditions in process water for used diffusers
αSAE	Aeration efficiency in standard conditions in process water
αFSAE	Aeration efficiency in standard conditions in process water for used diffusers

Standard conditions: 20°C, 1 atm, zero salinity, zero DO in water.

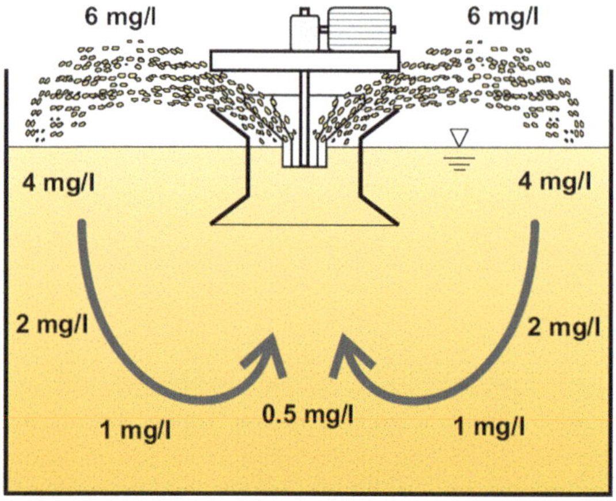

Figure 1.15 Schematic of DO patterns in tanks equipped with surface aerators.

Surface aerators move a liquid spray through air allowing the droplets to cool to the wet bulb temperature of the air (wet bulb temperature is a function of both the temperature and the humidity of air, so that cold, dry air would have the lowest wet bulb temperature) through evaporation. This cooling may be welcome when treating warm industrial effluents, or unwelcome when treating municipal wastewater in cold climates. The heat loss and the formation of aerosol spray are important factors and should be considered when selecting aeration systems. Issues such as spray drift and odour generation are especially important in urban areas.

Surface aerators are provided in two different configurations: high speed (i.e., direct drive), and low speed (i.e., with a gearbox). High-speed aerators typically rotate at 900–1200 rpm, and owing to the absence of the gearbox they are easy to install and are less expensive. On the other hand, the plume they generate is highly turbulent, which results in higher aerosol formation and potential floc shear. In general, high-speed aerators have lower SAE than low-speed aerators (0.9–1.3 kg O_2 kWh^{-1} for high speed, 1.5–2.1 kg O_2 kWh^{-1} for low speed).

The introduction of a gearbox between the electric motor and the impeller allows the aerator to rotate at lower speed (30–60 rpm). This is usually associated with increased capital costs and prolonged time for procurement (typically the gearbox is manufactured only after a purchase order is issued). The higher initial cost and procurement time may be partially compensated by higher AE.

Surface aerators cannot be used in deep tanks or lagoons without provision for mixing the lower sections, such as using draft tubes or lower impellers. A draft tube ensures that the upward pumping effect of the surface aerator extends to the lower levels of the tank. A lower impeller is mounted on a long shaft that

Figure 1.16 A low-speed surface aerator during installation (left) and in operation (right). The impeller at the bottom of the tank is used to ensure that solids and DO are distributed through the entire depth. The four structural beams act as support for the aerator as well as baffles to prevent vortex formation.

positions the impeller approximately 1 m above the tank bottom (Figure 1.16). High-speed and low-speed aerators are seldom used in tanks deeper than 4 m and 5 m, respectively, without draft tubes or lower mixers.

In the case of lagoons or earthen bottom tanks, the bottom must be protected from surface scour from the liquid. It is not uncommon for surface aerators to create depressions in earthen bottoms that may allow rocks and debris to enter and damage the propellers, or create structural instabilities. Additionally, for very shallow and wide lagoons, the aerator zone of influence may not extend to the edge of the lagoon, potentially causing the formation of mounds of solids in parts of the lagoon.

A type of surface aerator that provides aeration and mixing, as well as imparting horizontal velocity to the water, is the surface brush aerator or rotor, typically found in oxidation ditches (Figure 1.17). These low-speed surface aerators generally have elevated specific energy requirements (in other words, low SAE) because they pump the fluid as well as aerate. In oxidation ditches a minimum liquid velocity is required to keep the mixed liquor in suspension. In plug-flow tanks there is no need to maintain a liquid velocity. Since water is about three orders of magnitude denser than air, much of the energy of brush aerators is used for pumping liquid as opposed for aeration. Therefore, ditches with surface brushes are often candidates for retrofitting to fine-bubble diffusers and low-power subsurface mechanical mixers/pumps (such as 'banana-blade' mixers), which may significantly reduce the energy consumption of oxidation ditches.

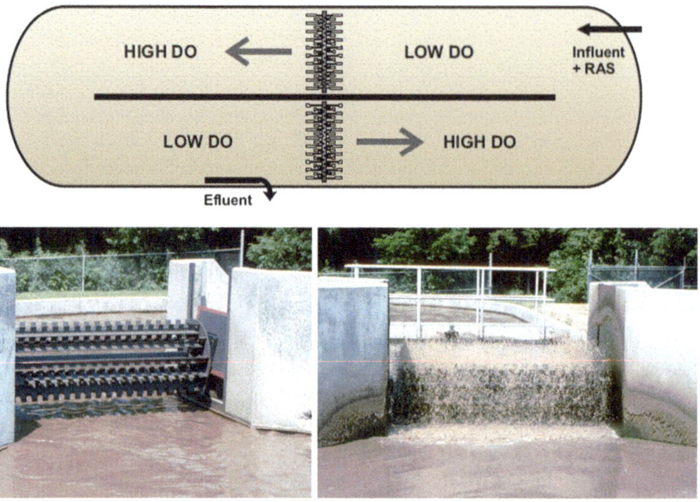

Figure 1.17 An oxidation ditch (schematic on top) equipped with surface brush aerator during standby (bottom left) and in operation (bottom right). The aerator's shaft is mounted on rails that can adjust the power drawn, lift the entire aerator above the free water surface, thereby avoiding any obstruction to liquid flow.

A less frequent application of surface brush aerators is shallow aerated lagoons (Figure 1.18). The surface brush is then mounted on a floating barge, open at the centre, where the surface aerator carries on mixing and aeration. This type of aeration technology is chosen in lagoons for ease of operation and maintenance,

Figure 1.18 A surface brush aerator mounted on a barge in an aerated lagoon.

as the barge can be easily tugged ashore for maintenance or repair. Also the aerator can be moved around to prevent the accumulation of sludge on the bottom.

1.3.2 Bubble aeration

Bubbles are typically released at full depth and transfer oxygen as they travel to the surface. The bubbles also induce liquid movement that contributes to mixing. In fact, since four-fifths of the air is nitrogen, we can conclude that at least 80% of the energy invested in bubble aeration contributes solely to mixing.

1.3.2.1 Coarse-bubble systems

Coarse-bubble systems all utilize relatively large orifices or slots to release air bubbles generally of dimensions above 50 mm. Bubbles in that range of diameters released in water cannot exist as spheres, but as spherical caps (which resemble the shape of jellyfish), because water is too thin. Coarse-bubbles are turbulent in nature and are characterised by a less severe surfactant interfacial accumulation, in part due to their high rate of surface renewal. Therefore, they have higher α factors when compared to fine-bubble systems (Eckenfelder & Ford, 1968; Kessener & Ribbius, 1934; Rosso & Stenstrom, 2006). Coarse-bubble systems are generally installed in full-floor configuration, to optimize air distribution and efficiency. In former times with less expensive energy costs, coarse bubble diffusers were often installed in a single row on the sides of plug flow tanks (spiral roll) or in two or more rows (cross roll, ridge and furrow). Increasing airflows in these diffuser arrangements typically increases the bulk liquid velocity but with only a limited impact on OTR. These systems required fewer diffusers, reducing capital cost, but are among the lowest-efficiency aeration systems, and are good candidates for upgrading.

Coarse-bubble diffusers are seldom affected by fouling or scaling. This is due to the large dimension and high turbulence of the discharge orifices, which makes them practically difficult to clog. On the other hand, these diffusers are always characterised by low SAE (in the range of 0.6–1.5 kg O_2 kWh^{-1}), because large bubbles travel very rapidly through the water column, and have relatively low surface-to-volume ratio.

Figures 1.19 and 1.20 show two models of coarse-bubble diffusers commercially available. Figure 1.19 shows two views of a sparger. Spargers belong to the first generation of coarse-bubble diffusers and are essentially capped hollow metal (older) and plastic (newer) bolts with one or more air release holes. The air travels through the air manifold, the drop leg, and inside the sparger, finally to be released through the sparger holes placed under the metal cap that enhances bubble shearing and prevents air channelling (bubbles travelling upwards as pearls of a necklace). A simpler system to produce coarse bubbles with relatively high efficiency is to use a piping grid, usually plastic, with approximately 5 mm diameter holes spaced at less than 0.5 m intervals along the bottom of the pipes.

Figure 1.19 Example of an air sparger. On the left is the bottom, partially disassembled, to show the inner structure. On the right is the top of the rubber cap that acts as check valve.

The 'chicken-feeder' diffuser (Figure 1.20) is a newer technology of coarse-bubble diffusers, characterised by a broader range of gas flow rates. This diffuser has on its side orifices of two sizes that proportionally increase in diameter going from the top to the bottom of the diffuser lateral. When the airflow rate is low, only the top orifices release air, but as the airflow increases, the back pressure increases and air flows from the row of larger orifices. At highest airflow rates, the air is forced through the open slot, which releases large spherical caps ($>$100 mm).

Due to their ability to accommodate high airflows, coarse-bubble diffusers have an advantage of being able to provide extremely high OTR within a given tank

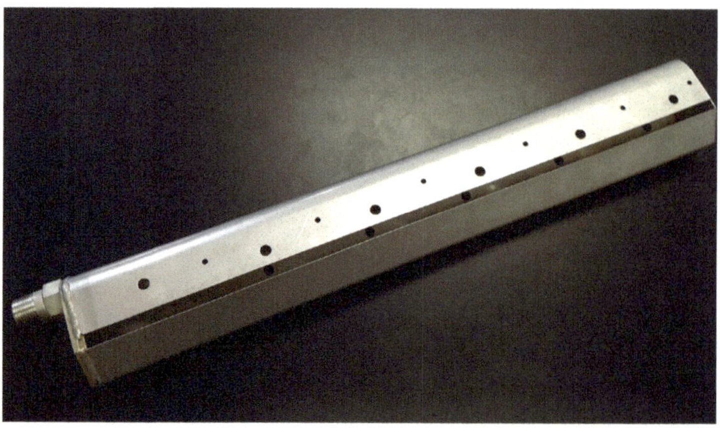

Figure 1.20 A 'chicken-feeder' coarse-bubble diffuser.

volume. Coarse-bubble systems generally have low SAE (SAE and OTR in general are inversely proportional). Obtaining large mass transfers (i.e., high OTR) requires high airflow rates that result in short bubble retention time and reduce the area for gas transfer. The maximum OTR of coarse-bubble diffusers can be several times higher than fine-pore or surface aerators and is usually limited by blower capacity (not the tank bottom surface area as in the case of fine-pore diffusers). Therefore, dense grids of coarse-bubble diffusers are sometimes the technology of choice for high-strength industrial wastewater treatment, where fine-bubble diffusers are a poor choice due to their limited air discharge and their extremely low alpha factors. For treatment systems not requiring high OTR per unit volume, such as municipal plants, coarse-bubble diffusers are a poor choice for energy conservation. Plants with coarse-bubble diffusers began to be replaced after the rapid rise in energy prices in the early 1970s, and municipal wastewater treatment plants most often use fine-pore diffusers.

1.3.2.2 Fine-bubble systems

Fine-bubbles can be produced by different technologies, by either releasing air through a porous plate, or by mechanically shearing large bubbles into small ones. Thc latter technology employs submerged turbines or jet diffusers, which create fine bubbles but do so without small orifices, and in both cases mechanical energy is used to shear large bubbles into fine bubbles. Fine bubbles from turbines or jets always have lower SAE (in the range of 1.2–1.8 kg O_2 kWh^{-1}) than fine bubbles from fine-pore diffusers. Also, their bubbles are released in few concentrated points, thereby often leading to uneven DO distribution and lower overall efficiency.

An important subset of fine-bubble systems is fine-pore diffusers; fine-pore diffusers produce their small bubbles by releasing compressed air through small orifices or pores in either punched membranes or porous material, such as ceramic stones or sintered plastic (Figures 1.21–1.24). Fine-pore diffusers are now the most commonly used diffusers in wastewater treatment in the United States and Europe. They have higher SAE (in the range of 3.6–4.8 kg O_2 kWh^{-1}), and are routinely used in full-floor configurations, taking maximum advantage of their efficiency. Fine-pore diffuser systems limit stripping of volatile organic compounds by virtue of their increased efficiency, which results in lower airflow rates (Hsieh *et al.*, 1993a, b). For the same reason, fine-pore diffusers also have reduced heat losses (Lippi *et al.*, 2009; Talati & Stenstrom, 1990).

Fine-pore diffusers have two important disadvantages: the first is the need for periodic cleaning. Every fine-pore diffuser fouls – *inevitably and inexorably*. More, inorganic scaling and material degradation (i.e., ageing) add complexity to the problem of quantifying cleaning and replacement schedules. No diffuser is immune from a combination of these phenomena, the difference is only in their extent for a particular installation. Entire sections of this book are dedicated to the systematic study of fouling, thus we will not elaborate much in this chapter.

Figure 1.21 The first-generation of fine-pore diffusers: ceramic domes. Air is fed through a hollow bolt, filling the space enclosed by the dome, and is then released through the pores of the sintered ceramic. The photograph on the right shows a diffuser harvested after prolonged time in operation, with visible biofouling on the outer surface. The coloured section of the dome interior shows the position of the bolt's air release hole, and is due to contaminants leaks in the air system.

The second important disadvantage of fine-bubbles (this one also applies to turbines and jets) is the large negative effect on transfer efficiency from wastewater contaminants, which is most often quantified by the α factor (ratio of process water to clean water mass transfer coefficients, or k_La_{pw}/k_La_{cw}).

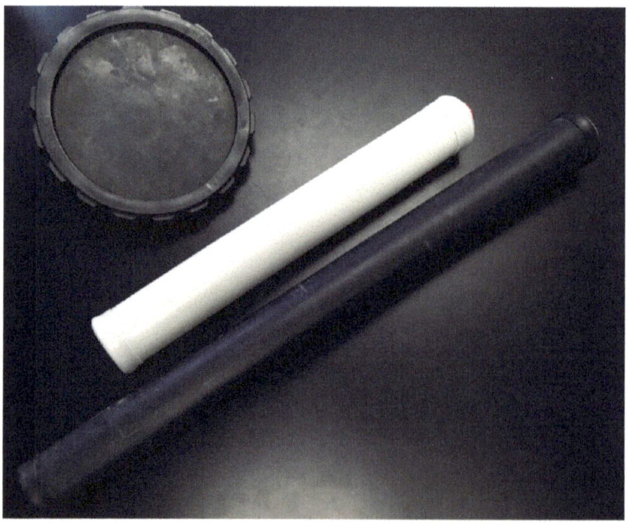

Figure 1.22 Assortment of diffuser models from different manufacturers. From the left of the photograph: EPDM membrane disc, silicone membrane tube and EPDM membrane tube.

Figure 1.23 Examples of full-floor installations of fine-pore diffusers. Left: ceramic discs; centre; polyurethane strips; right; polyurethane panels.

Fine-pore diffusers generally have lower α factors than coarse-bubble diffusers or surface aerators for similar conditions (Kessener & Ribbius, 1934; Stenstrom & Gilbert, 1981). Prior to the 1980s, many plants were designed with a constant α value of 0.8, which was considered as 'universal' for all types of aeration systems and for all conditions. In similar cases, nowadays we still witness poor design procedures employing such elevated and outdated values for α. It has been shown that different aeration methods have different α, and for fine-pore diffusers the initial α decreases over time in operation due to fouling or scaling (Rosso & Stenstrom, 2006). Furthermore, for fine-bubble systems α is a function of process conditions such as the mean cell retention time (MCRT) or the airflow rate

Figure 1.24 Fine-pore diffuser application to an aerated lagoon. The air pipes (on top of the diffuser unit) distribute the air along the lagoon, and vertical hoses convey the air from the air pipes to the diffuser units (placed close to the lagoon bottom). The image on the left shows one diffuser unit at a trade show.

(Rosso *et al.*, 2005). Finally, diurnal variations in load cause α to behave dynamically (Jiang *et al.*, 2017; Leu *et al.*, 2009).

Throughout this book we will elaborate on the virtues and shortcomings of different diffuser selections. The images in this section show different technologies, associated with different applicability. For example, when aggressive chemicals are present in the water, ceramic diffusers become the prime candidates. When energy efficiency is paramount, the highest-efficiency membrane diffusers in full-floor configuration are selected. For swing zones, membrane diffusers are selected due to the intermittent nature of operations. These are just some of the considerations, and an experienced engineer should always consider all the dimensions (in parallel manner to the blower selection) when selecting diffusers.

1.3.3 Comparative performance summary

The summary table compiles the authors' experience spanning the past four decades in many WRRFs (Table 1.3). The numbers are for relative comparison and should not be used for design purposes. Notice that fine-pore diffusers have the potential of achieving the highest AE, but give their best performance in processes with longer sludge retention time. This indicates that the efficiency is a function not only of the aeration system, but also of the process where it is installed (we will address this later in the book). Hence, our cautionary statement against using these data for design, since the numbers here may correspond to process configurations very different from the case under design. In the following chapters we will also explore the effects of process conditions on aeration performance.

Table 1.3 Summary of aeration efficiency (AE), standard aeration efficiency (SAE), and airflow weighted α factors for commercially available aeration systems. These ranges are the result of decades of field measurements by the authoring team in a large number of WRRFs operating under very different configurations. Use at your own peril!

Aerator Type	SAE kgO_2/kW-h (lbO_2/hp-h)	Low SRT AE at 2 mg/L DO	High SRT AE at 2 mg/L DO	Alpha Factor
High speed	0.9–1.3 (*1.5–2.2*)	0.4–0.8 (*0.7–1.4*)		0.47–0.64
Low speed	1.5–2.1 (*2.5–3.5*)	0.7–1.5 (*1.2–2.5*)		0.48–0.71
Turbine	1.2–1.8 (*2.0–3.0*)	0.4–0.6 (*0.6–0.9*)	0.6–0.8 (*0.9–1.4*)	0.30–0.47
Jets	1.5–1.7 (*2.5–2.8*)	0.8–1.2 (*1.3–2.0*)		0.52–0.71
Coarse bubble	0.6–1.5 (*1.0–2.5*)	0.3–0.7 (*0.5–1.2*)	0.4–0.9 (*0.6–1.6*)	0.50–0.64
Fine pore	3.6–4.8 (*6.0–8.0*)	0.7–1.0 (*1.2–1.6*)	2.0–2.6 (*3.3–4.4*)	0.20–0.55

1.4 REFERENCES

Eckenfelder W. and Ford D. (1968). Engineering aspects of surface aeration design. In: *Proceedings of 22nd Industrial Waste Conference*. Purdue University, Lafayette, Indiana, USA.

Hsieh C. C., Ro K. S. and Stenstrom M. K. (1993a). Estimating emissions of twenty VOCs: surface aeration. *Journal of Environmental Engineering*, **119**, 1077–1098.

Hsieh C. C., Babcock R. W. and Stenstrom M. K. (1993b). Estimating emissions of twenty VOCs: diffused aeration. *Journal of Environmental Engineering*, **119**, 1099–1118.

Jiang L. M., Garrido-Baserba M., Nolasco D., Al-Omari A., DeClippeleir H., Murthy S. and Rosso D. (2017). Modelling oxygen transfer using dynamic alpha factors. *Water Research*, **124**, 139–148.

Kessener H. J. N. H. and Ribbius F. J. (1934). Comparison of aeration systems for the activated sludge process. *Sewage Works Journal*, **6**, 423–443.

Leu S.-Y., Rosso D., Larson L. E. and Stenstrom M. K. (2009). Real-time aeration efficiency monitoring in the activated sludge process and methods to reduce energy consumption and operating costs. *Water Environment Research*, **81**, 2471–2481.

Lippi S., Rosso D., Lubello C., Canziani R. and Stenstrom M. K. (2009). Temperature modelling and prediction for activated sludge systems. *Water Science and Technology*, **59**, 125–131.

Loera J. (2013). *High-Speed Turbo Blowers Need ... Or Just Want [WWW Document]*. Wateronline. URL https://www.wateronline.com/doc/high-speed-turbo-blowers-need-or-just-want-0001 (last accessed 7 June 2018).

Rosso D. and Stenstrom M. K. (2006). Surfactant effects on α-factors in aeration systems. *Water Research*, **40**, 1397–1404.

Rosso D., Iranpour R. and Stenstrom M. K. (2005). Fifteen years of offgas transfer efficiency measurements on fine-pore aerators: key role of sludge age and normalized air flux. *Water Environment Research*, **77**, 266–273.

Stenstrom M. K. and Gilbert R. G. (1981). Effects of alpha, beta and theta factor upon the design, specification and operation of aeration systems. *Water Research*, **15**, 643–654.

Talati S. N. and Stenstrom M. K. (1990). Aeration-basin heat loss. *Journal of Environmental Engineering*, **116**, 70–86.

Chapter 2

Aeration fundamentals, performance and monitoring

Diego Rosso[1], *Michael K. Stenstrom*[2] *and Manel Garrido-Baserba*[1]

[1]*University of California, Irvine*
[2]*University of California, Los Angeles*

2.1 FUNDAMENTALS OF OXYGEN TRANSFER

Oxygen transfer is key to satisfy the microbial oxygen requirements by providing the required dissolved oxygen into the wastewater. Aeration systems transfer oxygen into the liquid media by releasing air through macroscopic orifices or porous materials, or by shearing the liquid surface with a mixer or turbine, or through direct contact of air and a large water surface. Falling droplets and rising coarse bubbles have large interfacial gas–liquid velocity gradients and can be grouped as high-flow regime interfaces, whereas fine bubbles have low interfacial velocity gradients and can be grouped as low-flow regime interfaces (Rosso & Stenstrom, 2006b; Rosso *et al.*, 2012).

When analysing or specifying aeration systems, it is important to define gas–liquid mass transfer efficiencies. These are necessary to compare different technologies, as well as to monitor aeration systems over extended time in operation. While some parameters can be calculated in clean water (i.e., absence of pollutants) others need to be corrected for wastewater (also called process water), as the effect of pollutants and process conditions exert a strong influence on mass transfer efficiency.

Editor: Diego Rosso
doi: 10.2166/9781780407845_0031

2.1.1 Clean water

First of all, *oxygen is a sparingly soluble gas*. This means that at any time we can use Henry's law to describe the equilibrium between atmospheric air and oxygen in the water. This will become apparent and useful when we talk about field conditions and DO saturation.

The target for aeration design is to provide oxygen to the water in excess to the oxygen requirements. We need to measure how much oxygen is transferred in clean water conditions, and scale it to process water conditions. The first step regards the rate of oxygen transferred in clean water, defined as the oxygen transfer rate (OTR, $kgO_2\ h^{-1}$). The OTR quantifies the capacity of the aeration system, that is, the amount of oxygen it can supply to the water per unit time

$$\text{OTR} = k_{\text{L}}a \cdot (\text{DO}_{\text{sat}} - \text{DO}) \cdot \text{V} \tag{2.1}$$

where $k_{\text{L}}a$ is liquid-side mass transfer coefficient (h^{-1}); DO_{sat} is dissolved oxygen in water at saturation ($kgO_2\ m^{-3}$), also represented as C_r; DO is dissolved oxygen in water ($kgO_2\ m^{-3}$), also represented as C^*_∞; and V is water volume (m^3).

Under standard conditions (clean water; new diffusers, i.e., no fouling: $F = 1$; dissolved oxygen concentration = 0 mg/L; water temperature = 20°C; and standard atmospheric pressure: 101.3 kPa), the standard oxygen transfer rate, SOTR, can be derived from equation (2.8) below:

$$\text{SOTR} = k_{\text{L}}a_{20} * \text{C}^*_{\infty\,20} * V_{\text{Tk}} * 24/1000 \tag{2.2}$$

As can be seen in equation (2.1) (and in equation (2.3)) the rate of oxygen transfer is proportional to the difference between the existing DO concentration and the equilibrium concentration of the oxygen in the water. In other words, the greater the difference between the existing concentration and the equilibrium concentration, the greater the *driving force* for oxygen transfer through the film (or fluid layer between the liquid bulk and the air bubble). This is can also be expressed by the following equation:

$$\text{d}C/\text{d}t = k_{\text{L}}a^*(C_{\text{r}} - C^*_\infty)V \tag{2.3}$$

The nomenclature followed for the following equations is that of ASCE's *Standard for Measurement of Oxygen Transfer in Clean Water* (ASCE, 2007), where $\text{d}C/\text{d}t$ is the rate of change of oxygen concentration ($mg\ O_2/L\ h$) at temperature T, also referred to as oxygen mass transfer coefficient (h^{-1}); C^*_∞ is the steady-state DO saturation concentration attained at infinite time, at water temperature T, and field atmospheric pressure P_b, that is, the equilibrium concentration of the oxygen in the water (mg/L), and C_r is dissolved oxygen in water at saturation ($kgO_2\ m^{-3}$).

It has to be noted that OTR defines the capacity of the aeration system regardless of its efficiency in performing oxygen transfer; therefore it is necessary to define additional parameters. The most common descriptor of energy efficiency is the aeration efficiency (AE, $kgO_2\,kW^{-1}\,h^{-1}$) defined as:

$$AE = \frac{OTR}{P} \tag{2.4}$$

where P is the power drawn by the aeration system (kW).

There are choices for types of power draw or P used in the AE, SAE or αSAE terms in Table 2.1. Wire power, or the power actually consumed by the electric motors used for the aerators or blower, is the most frequent choice and includes all losses in the system. Brake power, which is the output of a motor or gearbox, is sometimes used. In rare cases, water power or delivered power is used, which refers to the power actually transferred to the liquid being aerated. It is important not to confused power types and to be consistent in the design process.

Wire power is most often used because it is the best predictor of the actual power consumption. To use wire power, the inefficiencies of the blower, motors, and gearboxes must be known or measured. Brake power may be more convenient to use if the motors are being specified separately. A significant source of error in aeration system design results from confusing power types. Unless otherwise specified, wire power will be used through this chapter.

Without entering the details of mass transfer fundamentals available in many textbooks, based on the Lewis and Whitman (1924) two-film theory, we will present here the main equations pertaining to aeration.

For subsurface aeration devices (such as fine- or coarse-bubble diffusers), the oxygen transfer efficiency (OTE, %) can be used, defined as follows:

$$OTE = \frac{OTR}{W_{O2}} \sim \frac{(O_{2,in} - O_{2,out})}{O_{2,in}} \tag{2.5}$$

with OTR from equation 2.1, W_{O2} is the mass flow of oxygen fed to the aeration tank ($kgO_2\,h^{-1}$), and $O_{2,in}$ and $O_{2,out}$ representing mass fluxes of oxygen in and out of the clean water volume.

OTE gives an absolute efficiency of operation, and is often more convenient since it allows comparison of the aeration system without the complicating issues surrounding the blowers. Blowers are usually provided by a different manufacturer or contractor, and using OTE and blower efficiency independently simplifies specification and design of the aeration system.

In order to avoid bias due to site-specific environmental and process conditions, standard conditions are used and are defined as zero DO, zero salinity, 20°C, 1 atm. Therefore, results are typically reported as Standard Oxygen Transfer Efficiency (SOTE, %), Standard Oxygen Transfer Rate (SOTR, $kgO_2\,h^{-1}$), or Standard Aeration Efficiency (SAE, $kgO_2\,kW^{-1}\,h^{-1}$).

Table 2.1 Summary of all definitions used to specify aeration systems.

Symbol	Definition	Remarks
W_{O2}	Oxygen mass flow fed to the aeration tank	
OTR	Oxygen transfer rate in water	$= k_L a\ (DO_{sat}\text{-}DO)\ V$
SOTR	Oxygen transfer rate in standard conditions in clean water	
OTE	Oxygen transfer efficiency in clean water	$\sim(O_{2,\ in}-O_{2,out})/O_{2,in}$
SOTE	Oxygen transfer efficiency in standard conditions in clean water	
AE	Aeration efficiency in clean water	$= OTR/P$
SAE	Aeration efficiency in standard conditions in clean water	
$k_L a$	Liquid-side mass transfer coefficient	Measured in clean water tests
α	Alpha factor, that is, ratio of process- to clean- water mass transfer	$= \alpha SOTE/SOTE$
		$= k_L a_{process\ water}/k_L a_{clean\ water}$
F	Fouling factor	$= \alpha SOTE_{new\ diffuser}/\alpha SOTE_{used\ diffuser}$
αF	Alpha factor for used diffusers	$= \alpha\ F$
αSOTE	Oxygen transfer efficiency in standard conditions in process water	
αFSOTE	Oxygen transfer efficiency in standard conditions in process water for used diffusers	
αSAE	Aeration efficiency in standard conditions in process water	
αFSAE	Aeration efficiency in standard conditions in process water for used diffusers	

Standard conditions are defined as 20°C, 1 atm, zero salinity, zero DO in water. Key: *P*, power drawn; *V*, water volume.

2.1.2 In process water

Translating standard conditions to process conditions requires the use of several site-specific, empirical parameters. Process water (or wastewater) is characterized by dissolved and suspended contaminants, plus the presence of biomass, which cause a deviation in aerator performance from clean-water conditions. The parameter with the greatest impact is the α factor (see Sec. 2.1.3), which is defined as the ratio of the process water to clean water mass transfer coefficients, or

$$\alpha = \frac{(k_{\mathrm{L}}a)_{\text{process water}}}{(k_{\mathrm{L}}a)_{\text{clean water}}} \tag{2.6}$$

or

$$\alpha = \frac{\alpha\text{SOTE}}{\text{SOTE}} \tag{2.7}$$

where SOTE is oxygen transfer efficiency at standard conditions (%), and αSOTE is the oxygen transfer efficiency in process water at standard conditions except for the effect of contaminants on the mass transfer coefficient (%).

Note that we called alpha a parameter, while we have now recognised its variable nature. Later in the book we will elaborate more on the dynamic nature of alpha.

SOTE has to be measured under clean water conditions and following the guidelines of the ASCE standard for *Measurement of Oxygen Transfer in Clean Water* (ASCE, 2007) (see Sec. 2.3.2). This value, or an estimation of the real value, should be provided by the manufacturer of the aeration system (i.e., diffusers). On the other hand, αSOTE can be only measured *in situ* or in a sidestream setup (see Sec. 2.3.3).

Methods for translating standard to process conditions for all situations (i.e., barometric pressure, temperature, etc.) are required. There are a number of definitions used for describing aeration systems, as listed in Table 2.1 and used throughout this chapter.

Once the alpha factor is known, the OTR under field conditions can be calculated. The $k_{\mathrm{L}}a$ needs to be corrected for water temperature (using an Arrhenius correction), for process (i.e., dirty) water (using the alpha factor) and for diffuser fouling conditions (i.e., the F factor). Similarly, C^*_∞ must be corrected for water temperature, pressure, and process conditions.

The resulting OTR formula for field and standard conditions (αSOTR) is as follows:

$$\text{OTR} = \alpha F * k_{\mathrm{L}}a_{20} * \theta^{(T-20)}(\beta\,\Omega\,\tau C^*_{\infty\,20} - C_{\mathrm{r}}) * V_{\mathrm{Tk}} * 24/1000 \tag{2.8}$$

where:

OTR is the oxygen transfer rate at temperature T, ambient pressure p_{b}, with process (dirty) wastewater, and residual DO concentration C_{r} (kg O_2/d).

α is the ratio of the process water k_La of a new diffuser to the clean water k_La of a new diffuser, also known as the "alpha factor" (dimensionless).

F is the ratio of the process-water k_La of a diffuser after a given time in service to the process-water k_La of a new diffuser (no fouling at t_0: $F = 1$), also known as the "fouling factor" (dimensionless).

k_La_{20} is the apparent volumetric oxygen mass transfer coefficient in clean water at temperature $T = 20°C$, also referred to as oxygen mass transfer coefficient (h^{-1}).

θ is the Arrhenius temperature coefficient with a typical value of 1.024.

T is the mixed liquor temperature (°C).

β is a correction factor for dissolved solids and can be assumed to be in the range of 0.95–1.0 if the plant treats domestic wastewater. For wastewater with a considerable industrial component, its value can be calculated based on the dissolved solids content of the process water. In this case, β should be calculated as the ratio of the DO surface saturation concentration in water with dissolved solids equal to that of the process water, to the DO surface saturation concentration in clean water (both values can be found in tables).

Ω is a correction factor for atmospheric pressure and at tank depths of less than 6 m it may be approximated by p_b/p_s, where p_b is ambient barometric pressure and p_s is standard pressure at mean sea level.

τ is a correction factor for temperature and can be calculated based on published DO surface saturation values as

$$\tau = C^*_S / C^*_{S,20} \tag{2.9}$$

where C^*_S is a tabular value for DO surface saturation concentration at water temperature T (i.e., that recorded during the test), standard atmospheric pressure p_s, and 100% relative humidity; and $C^*_{S,20}$ is a tabular value for DO surface saturation concentration at water temperature 20°C, standard atmospheric pressure p_s, and 100% relative humidity ($C^*_{S,20} = 9.08$ mg/L, Metcalf & Eddy, 2014).

Another parameter that is currently used in aeration modeling is the standard oxygen transfer efficiency (SOTE), defined as the mass of oxygen transferred to the mixed liquor per mass of oxygen supplied to the tank (all under standard conditions defined in previous paragraphs). The SOTE is calculated as follows:

$$\text{SOTE} = \text{SOTR}/W_{O_2} \tag{2.10}$$

where

$$W_{O_2} = O_2 \text{ mass flow} = Q_{AIR} * \delta_{AIR} * O_2/\text{air} \tag{2.11}$$

and SOTR is standard oxygen transfer rate (kgO_2/d); SOTE is standard oxygen transfer efficiency (%); W_{O_2} is mass of oxygen fed to the aeration basin (kgO_2/d);

Q_{AIR} is airflow rate (m^3/d); δ_{AIR} is air density at standard conditions, approximately 1.2 kg/m^3; and O_2/air is the oxygen content of air, approximately 0.21 kg O_2/kg air.

Similarly, the oxygen transfer efficiency (OTE) under field or process conditions can be calculated as follows:

$$\mathrm{OTE} = \mathrm{OTR}/W_{O_2} \tag{2.12}$$

The relationship between k_La_{20} and SOTE can be derived from equations (9, 10 and 11). While the DO remains unchanged, the oxygen uptake rate OUR can be calculated directly as:

$$\mathrm{OUR} = \mathrm{OTR/V} \tag{2.13}$$

For the purpose of testing full-scale aeration systems using the off-gas method (see Sec. 2.3.3), the DO remains practically constant during the few minutes of a measurement, which allows for the direct calculation of the real-time *in situ* OUR.

As can be inferred from equation (2.8) the alpha factor and the fouling factor (which have a considerable variability), exert a tremendous influence on the oxygen transfer rate (OTR) and the oxygen transfer efficiency (OTE) under field conditions. The following sections describe the alpha factor and its influence on process design, equipment selection, operation and maintenance strategies, and energy costing aspects in treatment plants.

2.1.3 The mysterious alpha factor

The reduction in the oxygen transfer rate caused by contaminants present in wastewater is governed by the obscure alpha factor (α), as can be found in one of the earliest works on the subject by Groves *et al.* (1992). This particular label for an efficiency indicator is given by its unknown exact nature. The α factor is, then, reported as the most uncertain aeration process parameter (Karpinska & Bridgeman, 2016) since it presents a high variability and can be hardly predicted. The α factor is highly dynamic (Jiang *et al.*, 2017; Leu *et al.*, 2009) and cannot predict with certainty the aeration efficiency under different process conditions, either in a single facility (Amerlinck *et al.*, 2016) or across different facilities (Gillot & Héduit, 2008; Groves *et al.*, 1992).

Although, it is generally assumed that the most significant component affecting the α factor is surface-active agents or surfactants (see Sec. 2.1.3.1), the alpha factor is actually a composite parameter encompassing a wide range of potential components and conditions, all affecting in different ways and magnitudes. Some of the main causes leading to the reduction of aeration efficiency in wastewater range from chemical compounds (surfactants, organic substrates, etc.) to physical constraints (bubble coalescence, hydrodynamics, etc.), including the influence of microbial activity. Sludge rheology is particularly important in highly concentrated solid suspensions, such as MBRs or aerobic digesters.

The α-factor results are unknown *a priori* and may only be calculated *a posteriori*, for the wastewater composition and exact reactor conditions cannot be known beforehand. Lower-flow regime gas–liquid interfaces (such as the ones produced by fine-pore diffusers) generally have lower α than higher-flow regime interfaces (such as the ones produced by coarse-bubble diffusers or surface aerators) for similar conditions (Stenstrom & Gilbert, 1981). However, one must remember that high alphas are associated with lower SOTE and *vice versa*, due to the geometry and hydrodynamics of bubbles and droplets.

Differences in α amongst aeration systems were noted as early as the 1930s (Kessener & Ribbius, 1934), but were generally forgotten until the energy crisis of the 1970s increased awareness of the need for energy-efficient technologies. Prior to the 1980s, many plants were designed with an α of 0.8, which was considered as a "universal" α for all types of aeration systems. It has been shown that different aeration methods have different α, and for fine-pore diffusers the initial α decreases over time in operation due to fouling or scaling (Boyle & Redmon, 1983). Furthermore, for fine-bubble systems in activated sludge, α is a function of process conditions such as the mean cell retention time (MCRT) or the airflow rate (Rosso & Stenstrom, 2005). More, the effects of biomass on α has been documented (Cornel *et al.*, 2003; Germain *et al.*, 2007; Henkel *et al.*, 2011; Krampe & Krauth, 2003), providing the experimental evidence that is driving the efforts in studying the fluid dynamics of aerated tanks (e.g., Fayolle *et al.*, 2007).

2.1.3.1 Surfactant effects

Mass transfer depression caused by contaminants has long been observed (Mancy & Okun, 1960). Eckenfelder and Barnhart (1961) reported the effects of organic substances on mass transfer, showing that contamination as low as 15 mg/l of sodium lauryl sulphate (the main ingredient in soaps and detergents) can reduce mass transfer coefficients to 0.5 times the value in clean water. Considering the effects of surface-active agents (surfactants) on mass transfer Eckenfelder and Ford (1968) hypothesised that the interfacial accumulation of surface-active agents prevented oxygen molecules from exiting the bubbles because of "molecular obstruction". In fact, the interfacial accumulation of surfactants causes an increase in interfacial rigidity (hence in the drag coefficient), the reduction of internal gas circulation, and interfacial renewal rates (Rosso & Stenstrom, 2006b).

Conventional treatment, typically performed at lower MCRT, has lower biomass concentrations, and less chance for the dissolved substrate to be quickly adsorbed by the biomass. Higher MCRT operations have the advantage of higher biomass concentration. Given the same average MCRT, treatment systems using anaerobic selectors or coupling nitrification and denitrification have the additional advantage of partial removal or sorption of readily biodegradable substrate (rbCOD) in the selector zone. This is beneficial because of the decreased overall oxygen requirements (in the case of anoxic or nitrate-reducing selectors), and the decreased rbCOD accumulation at bubble surfaces that severely reduces oxygen transfer (Eckenfelder & Ford, 1968; Rosso & Stenstrom, 2006a).

The rbCOD is partially composed of surfactants, which are typically discharged as oils, soaps and detergents. The surfactants, because of their amphiphilic nature, accumulate at the air–water interface of bubbles. The surfactant accumulation increases the rigidity of the interface and reduces internal gas circulation and overall transfer rate (Rosso & Stenstrom, 2006b). Until recently, the intuitive concept of "molecular obstruction" was usually considered the cause of mass transfer depression. This phenomenon is dominant for stagnant gas–liquid interfaces, with zero interfacial fluid velocity, when molecular diffusion through the stagnant film is the only transport mechanism, and surfactant molecules in elevated concentrations can obstruct the molecular diffusion of oxygen molecules through the interface. For moving interfaces, turbulent transport towards the interface is the driving force for mass transfer, for two reasons: interfacial renewal rates and actual area covered by the surfactant molecules.

Figure 2.1 contains photographs at 1/500″ of coarse and fine bubbles in a 50 mg_{SDS}/l solution, and shows the inner fluid dynamic patterns with illustrations. The photographs show that the large majority of bubbles have diameters less than 1 mm. Clean-water tests without surfactant in analogous conditions produced mean bubble diameters of about 4 mm, larger than any bubble in the left half of Figure 2.1. In the figure one fine and one coarse bubble are highlighted and their interior circulation pattern is sketched to the side. The accumulation of surfactant at the fine-bubble interface occurs to a larger extent than for coarse bubbles, as the hydraulic residence time is longer and surfactant

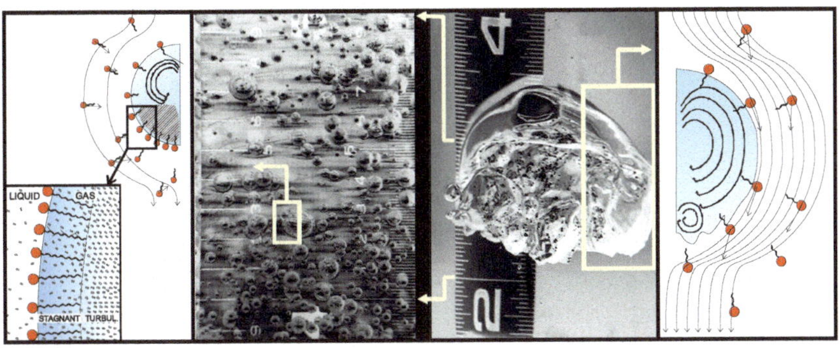

Figure 2.1 Comparison of fine (left photograph) and coarse (right photograph) bubbles generated by two different aerators operating at same airflow rate in the same surfactant solution. The outer illustrations show the different mechanism of interfacial accumulation in the two cases. The measuring scale is in inches (1 inch = 25.4 mm), and each scale sub-division is 1/10 in (= 2.54 mm).

molecules have longer time available for surface migration. Also, smaller bubbles have much lower interfacial velocity, and once the surfactants have attached to the surface, their hydrophobic tails inside the bubble reduce the internal gas circulation, acting like a baffle in a stirred reactor. Based on dimensional analysis results, fine bubbles covered with surfactant molecules have been shown to act like solid spheres, that is, with severely reduced internal gas circulation, which proved to be the principal cause for gas transfer depression (Rosso & Stenstrom, 2005).

The surfactant effect can be partially offset by increasing the flow regime (i.e., coarse bubbles). Figure 2.2 shows α as a function of the bubbles' Reynolds (Re) number, as originally hypothesized by Eckenfelder and Ford (1968) and as is represented in this figure by the curves. At low Re, an increase in Re decreases α. This occurs because the bubble is rising as a solid sphere. At higher Re, the buoyancy and drag forces are sufficient to cause internal bubble circulation. At very high Re, practically achievable only with coarse bubbles, surfactant effects are nearly offset, increasing α factor at the expense of energy efficiency or low SAE. Figure 2.2 also shows the impact of two different surfactants, a "fast" one with high diffusivity (sodium dodecyl sulphate, a.m.u $\sim 10^2$) and a "slow" one with lower diffusivity (polyvinylpyrrolidone, a.m.u $\sim 10^4$) surfactant. The fast surfactant more dramatically suppresses the transfer rate because of its greater diffusion rate and greater accumulation at the bubble surface.

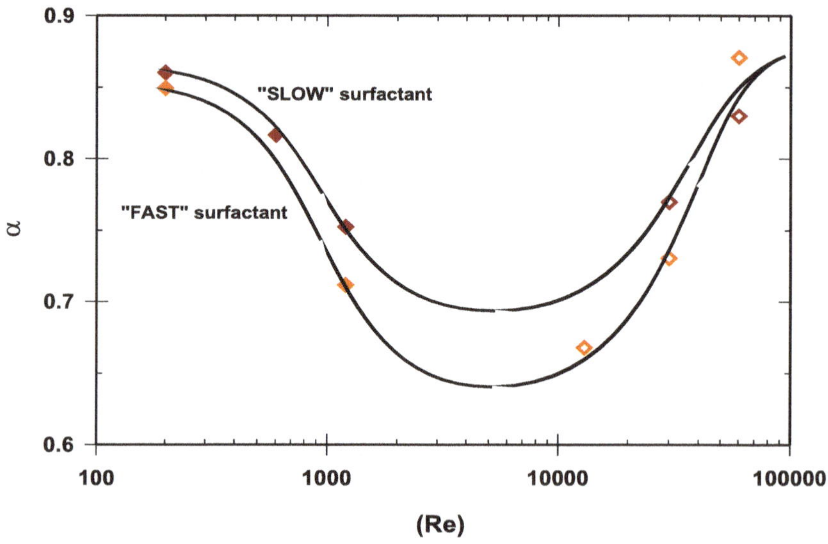

Figure 2.2 Effect of flow regime on α factors for fine (low Re) and coarse (high Re) bubbles in two different surfactant solutions. The experiments are from Rosso and Stenstrom (2006b) and the curves are an adaptation of Eckenfelder and Ford (1968).

The high turbulence associated with coarse-bubble aerators makes them behave more like surface aerators than fine-bubble aerators. Therefore, high-turbulence aerators may achieve better transfer rates, but at the expense of greater energy density, and lower aeration efficiency.

2.1.4 Fine bubbles, coarse bubbles, and droplets

2.1.4.1 Technologies for bubbles and droplets generation

Aeration systems transfer oxygen into a liquid media by either diffusing gas through a gas–liquid interface, or dissolving gas into the liquid solution using a semi-permeable membrane. In other words, environmental technologies can rely on: (1) interfacial gas transfer, with a gas–liquid interface created by either shearing the liquid surface with a mixer or turbine, or (2) by releasing air through spargers or porous materials. Moreover, a third category of technologies exists: (3) novel devices (referred to as bubbleless, e.g., the MABR) which utilize the intrinsic porosity of "true" semi-permeable membranes, such as reverse osmosis membranes, for transferring gas into the liquid. These membranes allow the transit of water and air across the membrane, without permitting the passage of solute or suspended matter and without generating an interface.

There exists other aeration equipment, such as submerged turbines or jet diffusers, which can create fine bubbles without using small orifices, and are fine-bubble but not fine-pore aerators. These technologies use mechanical energy to shear large bubbles into fine bubbles.

2.1.4.2 Bubble formation

In fine-pore diffused aeration, bubble formation is created by means of continuously pushing compressed air through diffusers with small orifices or porous materials. Bubble size depends on several factors such as flow rate, inlet pressure, and the contact angle with the rubber membrane. Among these factors, the flow rate has the largest effect on the bubble size, followed by membrane material and contact angle. The punch size has a moderate effect on the bubble size, whereas the punch length and punch direction have only a slight effect on bubble size. Finer bubbles are preferable to the coarse-bubble diffusers. Fine-pore diffusers provide larger surface area and longer residence time, which will improve the standard oxygen transfer. In addition to the above factors, wastewater characteristics and submergence must be taken into account under particular conditions (Wagner & Pöpel, 1998); the liquid depth will modify the bubble column residence time (Alkhalidi & Amano, 2015). A detailed description of bubble formation in submerged orifices, and a review of the theoretical models encompassed by this process, are provided by Clift *et al.* (2005).

Factors to consider in bubble formation may vary according to flow characteristics of the aerobic reactor of the activated sludge process. Important differences are found when comparing bubble formation near quiescent conditions, under a no horizontal flow regime with bubble formation under liquid cross-flow conditions (Fayolle *et al.*, 2010). In the latter case, the dynamics of bubble formation are more complex (velocity gradient, bubble inclination, and distortion). The aforementioned flow regimes are characteristics, in some variation, of the activated sludge process, as is the case of the sequencing batch reactors (SBR) as well as the oxidation ditch, respectively.

Under a no horizontal flow regime, the detached bubbles have significantly smaller sizes and higher frequencies when compared to bubble formation under quiescent liquid conditions. The bubble detachment is controlled by the drag force as a result of the liquid motion and not the buoyancy force (Loubière & Hébrard, 2004).

2.1.4.3 Limitations in assessing oxygen transfer rate

The oxygen transfer mechanism in the activated sludge process is still far from being fully described or modelled, despite the available knowledge. Nowadays experts acknowledge that the oxygen transfer rate can be evaluated by the overall volumetric oxygen transfer coefficient, K_La, which is composed of K_L, the overall oxygen transfer coefficient, and a, the interfacial area available for exchange. K_La is generally determined for clean water, following the ASCE standard. It is used as a constant or as a function of air flow rate, based on the assumptions that: (i) the aeration tanks are completely mixed; (ii) the rheological characteristics of the activated sludge do not strongly impact on the oxygen transfer and; (iii) aeration is homogeneously provided. These assumptions are known to limit the accuracy of the aeration efficiency models and as such hamper any optimization study making use of them (Amaral *et al.*, 2016; Karpinska & Bridgeman, 2016). Therefore, these aeration models have been extended using empirical relationships considering the aforementioned α-factor (see Sec. 2.1.3), to introduce the dependency on operational and environmental conditions of the process.

Moreover, bubble size distribution was found to play a role on the estimation of the oxygen transfer affecting K_L and a. In some studies, a single average bubble size is assumed (Fayolle *et al.*, 2007). These limitations are expected to be slowly overcome, however, to date only some studies (at lab scale and not representing real full-scale plants) are trying to go beyond these assumption through photographic quantification (McGinnis & Little, 2002), through the introduction of interaction models which are accounting for bubble coalescence and breakage (Karpinska & Bridgeman, 2016), or expanding the existing knowledge on the effects of viscosity (Amaral *et al.*, 2016) by using computational fluid dynamics (CFD).

2.1.4.4 Interfacial velocities

In wastewater treatment there are two regimes of interfacial velocity: high and low, which correspond to coarse-bubble and surface aerators, and fine-pore diffusers, respectively.

High interfacial velocities. Surface aerators shear the wastewater surface producing a spray of fine droplets that land on the wastewater surface within few seconds and over a radius of a few metres. Falling droplets and rising coarse bubbles have large interfacial gas–liquid velocity gradients and can be grouped as high-flow regime interfaces, whereas fine bubbles have low interfacial velocity gradients and can be grouped as low-flow regime interfaces. Coarse bubbles are larger, customarily produced by orifices 6 mm and larger (ASCE, 1991), and may have diameters as large as 50 mm. The high turbulence associated with coarse-bubble aerators makes them behave more like surface aerators than fine-bubble aerators. High-turbulence aerators may achieve better transfer rates, but at the expense of greater energy density, and lower aeration efficiency.

Low interfacial velocities. Diffusers are nozzles or porous surfaces placed on the tank bottom that release bubbles travelling towards the tank surface. In general, bubbles are considered fine when their diameters are less than 5 mm. Polymeric or ceramic porous sintered materials and polymeric punched membranes are usually grouped together and referred to as fine-pore diffusers. While turbines and jet aerators produce high-speed bubbles, once these reach their terminal velocity, their interfacial velocity is same as the same bubble originating from a fine-pore diffuser. Fine-bubble diffusers have greater mass transfer decline in process water than coarse-bubble or surface aerators.

2.1.4.5 Understanding differences in interfacial regimes

Turbulence controls gas transfer depression and can be characterized by the interfacial Reynolds or Péclet numbers and the gas transfer by the interfacial Sherwood number. The Sherwood number can be correlated to the Péclet number to account for turbulence and to a dimensionless number, called the interfacial contamination number, to account for interfacial contaminant accumulation (Rosso & Stenstrom, 2006b).

Figure 2.3 shows the evolution of mass transfer coefficients with increasing time. As time elapses, surfactant molecules migrate towards the surface, and hinder the interfacial renewal process, resulting in a rapid decrease in the volumetric mass transfer coefficient k_La, which is the product of liquid film coefficient, k_L, and interfacial area, a. Since bubbles are stable at smaller diameters in contaminated waters, the interfacial specific area is higher. Since k_La decreases, k_L must decrease more rapidly than the interfacial area (a) increases. Eckenfelder and Barnhart (1961) observed that k_L decreases with increasing surfactant concentrations, but k_La is minimum at low surfactant concentrations, and partially recovers at higher concentrations. This is the effect of a larger interfacial area (a), but reduced liquid film coefficient k_L. In the case of moving interfaces, as for diffused and surface aerators, turbulence exists behind the interfacial laminar films.

The Reynolds number is defined as $\mathrm{Re} = du/\nu$; where d is the characteristic length, u the velocity, and ν the kinematic viscosity. At a given interfacial flow

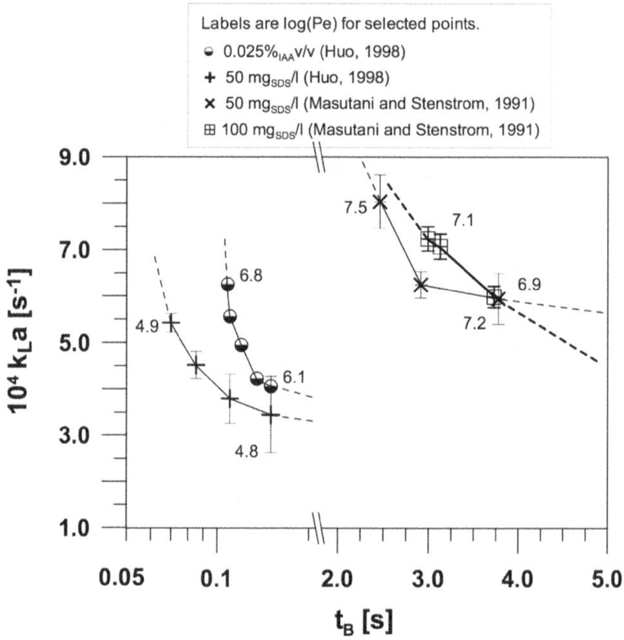

Figure 2.3 Evolution of mass transfer coefficients over increasing bubble surface age t_B. Labels are log (Pe) for selected points calculated at the interface, where liquid-side diffusivities are higher than bulk values.

regime, hence at a given interfacial (Re), the film renewal rate is decreased with increasing contamination, and lower internal gas circulation rates in bubbles. Roy and Duke (2004) photographed two-dimensional DO concentration gradients near surfaces contaminated with Triton X-100, showing reduced circulation outside contaminated bubbles, with lower interfacial O2 concentration gradients for higher contaminations. The intuitive concept of "molecular obstruction" is dominant for stagnant gas–liquid interfaces, with zero interfacial fluid velocity, when molecular diffusion through the stagnant film is the only transport mechanism. For moving interfaces, turbulent transport towards the interface is the driving force for mass transfer, for two reasons: interfacial renewal rates and actual area covered by the surfactant molecules.

Higher contaminations with lower surface tensions cause higher boundary layer thickness, creating lower probability for turbulent eddies to reach the interface and carry "fresh" bulk fluid gas packets, causing lower renewal rates. Figure 2.3 shows different mass transfer coefficients at same contaminations (50 mgSDS/l). The higher mass transfer coefficients from higher interfacial velocities (characterized by higher interfacial Péclet numbers) are shown in the same plot (Fig. 2.4). The same contaminations yield different mass transfer coefficients in different flow

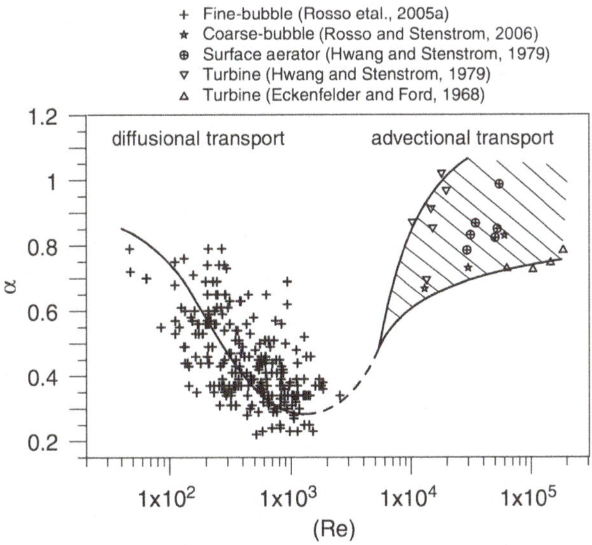

Figure 2.4 Alpha-factors at different interfacial flow regimes. The shaded area represents a region of possible trend lines parametric in the energy density. Adapted from Rosso and Stenstrom (2006b).

regimes; furthermore, higher interfacial velocity results in mass transfer coefficients in the same range.

2.2 FACTORS AFFECTING OXYGEN TRANSFER

Different factors impacts oxygen transfer in aeration systems, from environmental to process-related. In the following sections, the state-of-the-art regarding the factors affecting oxygen transfer is presented. For example, diffuser-related characteristics such as type, installation depth, distribution, time in operation and air flow rate will influence the capacity of the system to transfer air in the form of dissolved oxygen. Similarly, aerobic reactor features such as depth, volume, and type of reactor, as well as operating conditions such as mean cell retention time, nutrient removal processes, turbulences, wastewater composition and fouling will influence the oxygen mass transfer (Eckenfelder, 1959; Groves *et al.*, 1992; Henkel, 2010). Nevertheless, the efficiency can also be impacted by other factors that can be controlled or predicted as temperature, barometric pressure, salinity or water pressure.

2.2.1 The impact of sludge retention time

The most important process variable to affect aeration efficiency in activated sludge is the mean cell retention time. MCRT is directly related to the biomass concentration, and dictates oxygen requirements. Aeration efficiency and α factors are higher at higher MCRTs. Biological nutrient removal processes, by operating at increased MCRTs, have improved aeration efficiency. Furthermore, anoxic and anaerobic selectors in plants with nutrient removal have beneficial effects that go beyond nutrient removal or improved settling characteristics. By utilising the readily available carbon in the wastewater, they more rapidly remove surfactants, which have the most dramatic impact in reducing oxygen transfer.

Literature studies (Rosso & Stenstrom, 2005; USEPA, 1989) showed that the oxygen transfer efficiency is directly proportional to MCRT, inversely proportional to air flow rate per diffuser, and directly proportional to geometry parameters (diffuser submergence, number and surface area of diffusers). Figure 2.5 illustrates this concept. The MCRT determines the net oxygen requirement and relates to the degree of treatment and removal of oxygen transfer reducing contaminants. Higher MCRT systems remove or sorb the surfactants early in the process, which improves average oxygen transfer efficiency. The net effect of increasing MCRT is to increase the oxygen requirements, improve removal of biodegradable organics, and improve the overall oxygen transfer efficiency (Rosso & Stenstrom, 2007). The increase in oxygen requirement is partially or more than offset by the savings produced by the higher transfer efficiency, for processes in warm climates.

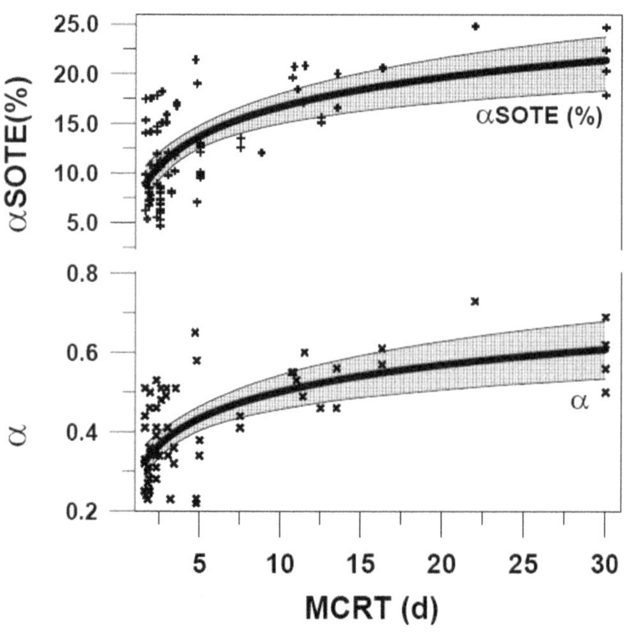

Figure 2.5 Effect of MCRT on αSOTE and α. Shaded areas are 95% confidence intervals.

In addition to these advantages, there is growing evidence that processes operating at higher MCRT are more efficient in removing anthropogenic compounds, such as pharmaceuticals (Göbel *et al.*, 2007; Soliman *et al.*, 2007). Studies reported removals up to 90% for the endocrine disruptor 17a-ethinylestradiol (EE3) after a wastewater treatment plant was retrofit to remove nutrients at MCRT of 11–13 days (Andersen *et al.*, 2003). Operation at longer MCRT in order to enhance removal of trace organics will become more important as wastewater water reclamation becomes more widely practiced.

2.2.2 Role of selectors

Almost all new activated sludge process designs utilize selectors, either anoxic or anaerobic. The benefit of selectors is the reduction of filamentous organisms (Harper & Jenkins, 2003), which improves SVI and reduces the probability of sludge bulking and rising sludge blankets in secondary clarifiers (Schuler & Jang, 2007). One study surveyed 21 plants with anoxic and anaerobic selectors, and reported that all plants showed improvement after selector installation (Parker *et al.*, 2003). Among the plants with anoxic selectors, 70% had sludge volume index (SVI) lower than 200 ml/g. Plants using anaerobic selectors were even better and more than 90% of the plants had SVIs less than 150 ml/g. Martins *et al.* (2004) reported similar results and concluded that better operation is achieved if there is a first anaerobic stage. The activity of phosphorus accumulating organisms (PAO) was reported to increase even when operating a strictly anoxic selector, with PAO improving the floc structure and biomass density (Tampus *et al.*, 2004).

An advantage of selectors is the removal or sorption of a fraction of the carbonaceous load, that is, the readily biodegradable COD (rbCOD). This is illustrated in Figure 2.6. The rbCOD is partially composed of surfactants, which are typically discharged as fatty acids, oils, soaps, and detergents. The surfactants, because of their amphiphilic nature, accumulate at the air-water interface of rising bubbles, reducing oxygen transfer efficiency. Removal of the rbCOD can improve oxygen transfer efficiency, which reduces operating costs for aeration (Rosso & Stenstrom, 2006a, 2007).

In Figure 2.7 we report the results of a survey of many WRRFs tested for oxygen transfer efficiency. The diffusers tested were of all makes and models and had been operating for a wide variety of times (from brand new to more than 5 years). In this figure we labelled the diffusers as NEW when they had been within 1 month from installation, USED when they had been between 2 and 24 months in operation, OLD when they had been more than 24 months in operation, and CLEANED for those that had been cleaned within a month. We clustered the data by process type, that is, BOD removal only (i.e., conventional), nitrification only, and NDN. Each point in the plot is the airflow weighted average of an off-gas test for a tank at a WRRF.

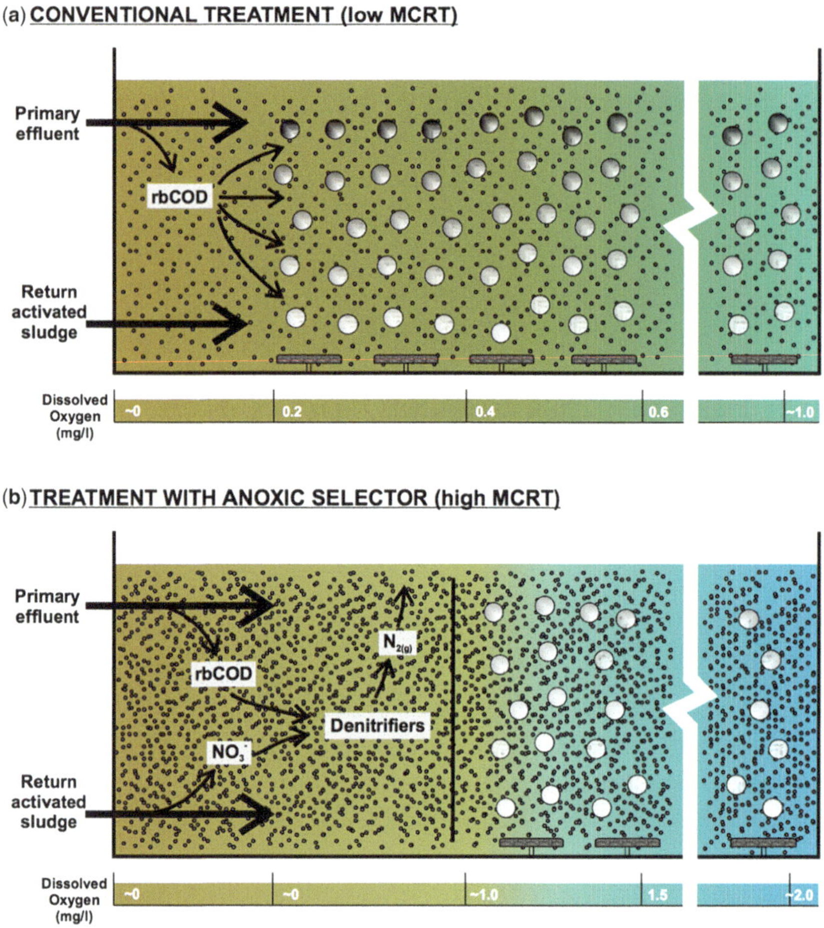

Figure 2.6 Schematic view of the role of selectors on the pathway and fate of readily biodegradable COD (containing surfactants). The graph shows proof of the different process efficiencies. The data points in the graph were measured in a treatment plant with two independent activated sludge trains with independent clarifiers, treating the same wastewater.

The first observation is that for each data cluster the spread is wide: alphas always range +/− 0.1, indicating a high variability amongst sites and times in operation. However, if we look at the averages and error bars there is a clear upwards trend. In fact, conventional processes have to be operated at the lower end of MCRT, while the others benefit from the longer sludge age. Also, the MCRT range is same for nitrification-only and NDN and the only difference appears to be the presence of anoxic selectors.

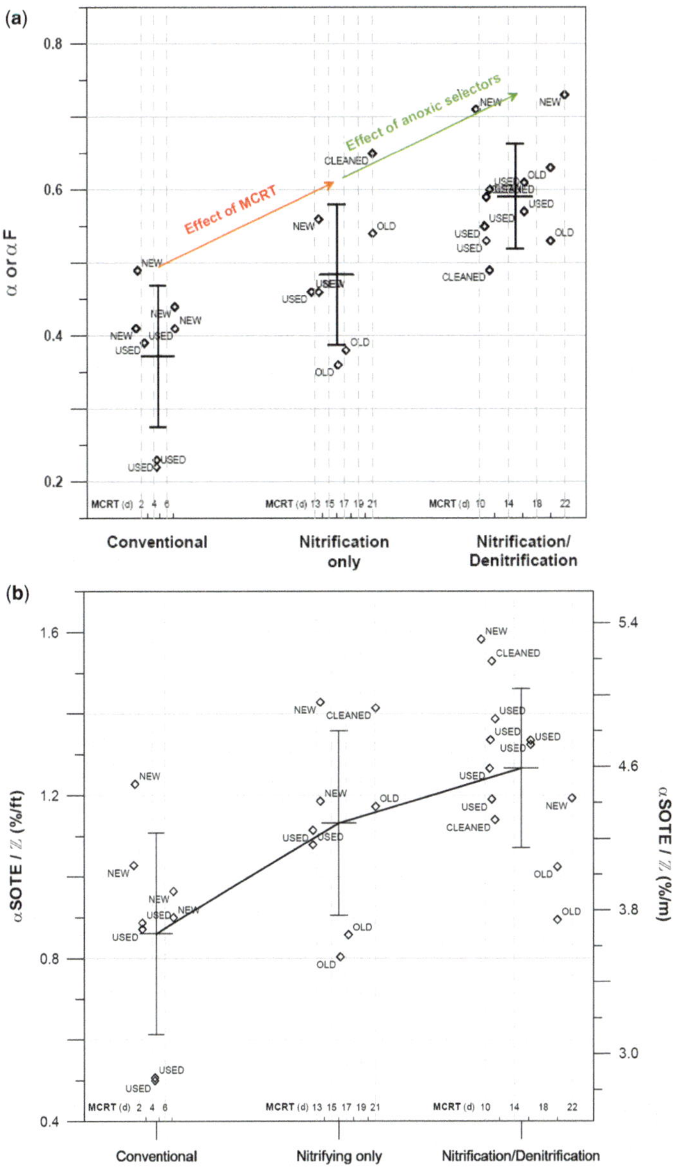

Figure 2.7 a) Normalized Standard Oxygen Transfer Efficiency for selected plants operating with different layouts; b) The evolution of α or αF factors for selected plants operating with different layouts. Labels refer to the diffuser status: NEW (within 1 month from installation), USED (between 1 and 24 months of operation), OLD (over 24 months in operation), and CLEANED (within 1 month from a cleaning event). Note the increase in α or αF with the increase of MCRT, and the further increase due to anoxic selectors (the average MCRT for N-only and NDN is the same).

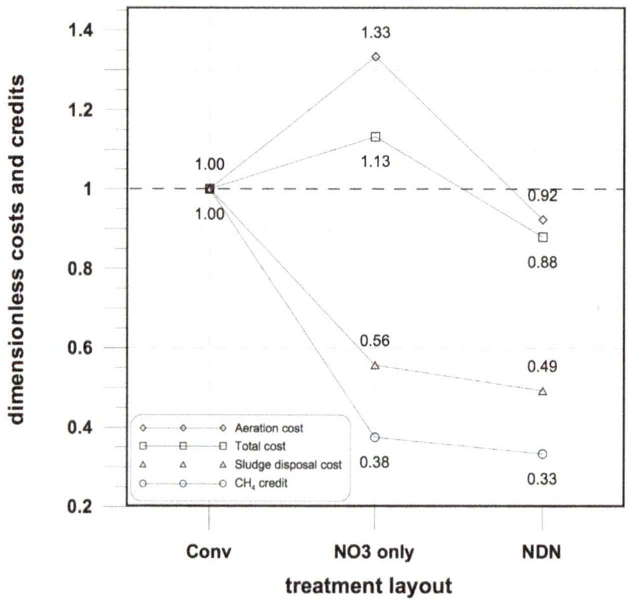

Figure 2.8 Dimensional and dimensionless costs and credits for different process layouts.

High-rate processes, characterised by very low sludge age (e.g., the A stage in the A/B process), may have a range of alpha factors that is not a downward linear extrapolation of the trend visible in Figure 2.7. Research on the subject is still in progress (Rosso & Al-Omari, 2018).

For these and other reasons, anoxic selectors for nitrification/denitrification (NDN) should always be evaluated as an alternative to conventional treatment. Our previous analysis (Figure 2.8) showed that NDN operations can have lower total operating cost (aeration + sludge disposal costs – methane credit) than a nitrifying-only or a conventional treatment plant (Rosso & Stenstrom, 2005). If conventional process operating cost was normalized to 1.00, nitrifying-only will have a total cost of 1.13, and NDN operations will have a total cost of 0.88. NDN operation offers an oxygen credit due to its process nature, and higher oxygen transfer efficiencies associated with the higher MCRT. The two factors overcome the additional oxygen demand that produced by the longer MCRT.

2.2.3 Reactor characteristics

2.2.3.1 Airflow rate

In fine-pore diffusers, the bubble size distribution is linked to the airflow rate, with higher flow per diffuser resulting in larger bubble diameters. Consequently, these larger bubbles result in a smaller interfacial area. Therefore, as the airflow rate per diffuser increases, the clean water performance SOTE decreases (US EPA, 1989). The cause for more airflow in a plant is an increase in oxygen demand load, which corresponds to higher concentrations of contaminants that depress the alpha factor. Hence, in process water, when higher airflow is required, the lower SOTE compounds with the alpha factor decreased by the more concentrated contaminants, and the overall result is an amplified decrease in αSOTE.

Furthermore, increased aeration rates generally result in increased OTRs, because of the effects of turbulence caused by aeration. Higher airflow rates and mixing turbulence can reduce the depth of the boundary layer, improving the oxygen transfer coefficient K_L and OTR (Ji & Zhou, 2006; Vogelaar *et al.*, 2000).

2.2.3.2 Diffuser density

Typically, an increase in diffuser density, defined as the area covered by the diffusers in relation to the total area of the tank floor, is expected to result in a higher OTE. However, a maximum value exists for the density of diffusers in which the SOTE increase is minimal (diminishing returns), which is determined by diffuser size, airflow rate, and the space between diffusers (US EPA, 1989).

Care must be applied during design to prevent an overzealous specification of diffusers. If too many diffusers are selected, based on a cost/benefit analysis that indicates that more diffusers would have an overall energy life-cycle cost, the risk of excessively low air flow per diffuser may be incurred. In the case when diffusers are operated at excessively low air flow, the bubble discharge may be uneven and the mixed liquor may no longer be well-mixed.

2.2.3.3 Flow regime

Each of the reactor configurations, whether plug flow or complete-mix, is accompanied by a particular set of conditions that affect the OTE. In the plug flow regime, loads are variable along the aerobic reactor and generally decrease from the reactor inlet to the outlet. Thus, the α-factor is low at the inlet of the aerobic reactor and high at the outlet. For a complete-mix aerobic reactor these variations are generally not significant. On one hand, complete-mix reactors offer uniform distribution but lower α factors, when compared with the profiles of plug flow reactors. On the other hand, mixed reactors typically exhibit higher average α factors than the influent end of a plug flow reactor. NDN processes with internal recycling (e.g., MLE processes) show profiles that lay in the middle between the two boundary conditions of plug flow and well-mixed (Tseng *et al.*, 2015).

A commonly encountered feature of plug-flow reactors is tapered aeration. In this configuration, the diffuser density is higher at the influent end of the tank than at the effluent end of the tank. Because the influent side of the reactor exhibits a higher load, the point of entry exhibits higher air demand. On average, half of the load of a plug-flow reactor is treated in the first 25–35% of the aerated process, if we observe profiles along the reactors.

Regarding stirred reactors, the effect of rotational speed, gas flow rate, and impeller type have been studied. In general terms, for all operating conditions including the stirred reactor, k_La values increase with increasing total power density. For the same total power consumption, the aeration efficiency values obtained with bubble columns are higher than those provided by the stirred gas–liquid reactor. This difference is explained by the higher values of interfacial area obtained in bubble columns per unit energy invested (Bouaifi & Roustan, 2001; Bouaifi *et al.*, 2001).

2.2.3.4 Depth of the aerobic reactor

The OTE of the aerobic reactor increases with increasing depth because of the longer residence time of the bubbles, and the larger partial pressure of oxygen at the moment of bubble formation. However, Wagner and Pöpel (1998) and Gillot and Héduit (2008) showed how the alpha factor decreases with increasing depth. As the partial pressure increases with diffuser depth, the operating pressure for the blower also increases. Nonetheless, the Standard Aeration Efficiency (SAE) remains constant because an increase in depth reflect as an increase in energy consumption (US EPA, 1989; Water Environment Federation, 2009). When measuring gas transfer coefficients in deep tank systems, the difference between the true k_La and the apparent k_La (i.e., the k_La actually measured) can be large. Researchers had discussed these phenomena and the corrections during the period leading to the compilation of the earlier testing standards (Downing & Boon, 1963). The apparent measurement of k_La reflects in an analogous departure in the calculation of alpha factors. The magnitude of the effects is usually small, and much less than the experimental error associated with field testing. However, the case of very deep tanks (>10 m) is still unresolved and *in situ* testing remains the only solution to quantify alpha factors at this time.

2.2.4 Diffuser fouling, scaling, and cleaning

The oxygen transfer efficiency of fine-pore diffusers inevitably decreases over time (Figs 2.11 and 2.12). At the same time, the back pressure (often called dynamic wet pressure or DWP) usually increases, and in some cases dramatically. This DWP increase is due to clogging of pores in ceramic and membrane diffusers (USEPA, 1989), or is associated with a permanent change in orifice characteristics for polymeric membranes (Kaliman *et al.*, 2008). Both effects account for the decrease in overall process efficiency and power wastage. Cleaning fine-pore diffusers is almost always required and restores process efficiency and reduces power costs (Fig. 2.9). Observations of 94 field tests show that efficiency decreases with time and the greatest rate of decrease occurs in the first 24 months of operation (Rosso & Stenstrom, 2006b). Efficiency decline was quantified and

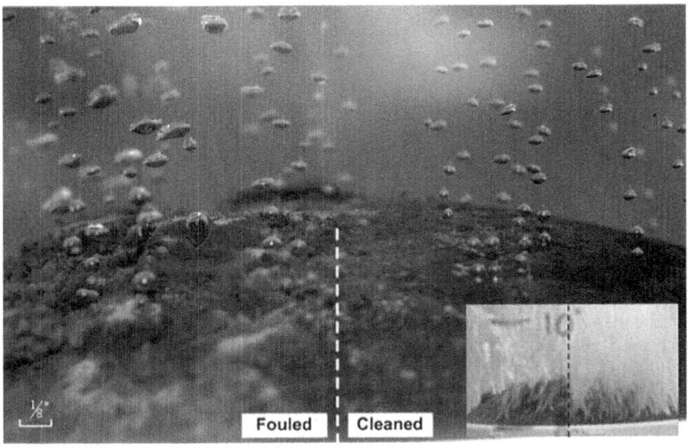

Figure 2.9 A fouled EPDM disc diffuser harvested from a conventionally operated treatment plant after 2 years in operation (MCRT <2 days), cleaned by surface brushing on its right-half. The difference in number and size of bubbles released is due to the biofouling layer still present on the left-half of the membrane. The air released from the orifices on the left-half has to travel through the biological layer, and during this travel time it has the opportunity to coalesce and form fewer and larger bubbles with disadvantaged mass transfer. The detail shows the membrane at operating air flow rate. (Large photo courtesy of Shao-Yuan Ben Leu).

included in cost analyses, and the net-present worth was compared to cleaning costs. The cleaning frequency is always higher for higher fouling rates and optimal frequency was as short as 9 months and never more than 24 months.

Due to the chemical nature and the morphology of these materials, they experience fouling and scaling depending on process conditions, water quality, diffuser type, and time in operation (US EPA, 1989). As a result, fine pore diffusers need to be routinely cleaned. The choice of cleaning frequency and method determines the long term efficiency and benefits of using fine pore aeration.

Different methods have been used to clean fine-pore diffusers and these methods vary in complexity and cost. The simplest method is to dewater the aeration tank and wash the diffusers from the tank top, using plant effluent (Fig. 2.10). This form of cleaning, called tank top hosing, is effective in removing biological slime accumulation, and generally partially or completely restores efficiency. For cases where inorganic precipitates (silica, calcium carbonate, gypsum, etc.) have caused scaling, acid cleaning may be required. Manually washing with low-strength hydrochloric acid (10 to 15% w/w) is popular and acid gas cleaning using hydrochloric acid gas or acetic acid injected to air distribution lines is also possible (Schmit *et al.*, 1989). This technique is limited to ceramic diffusers, due to the incompatibility between currently employed polymeric membranes and strong inorganic acids.

Figure 2.10 Examples of diffuser cleaning. Top: a dewatered tank equipped with ceramic domes before (left) and during (right) cleaning by means of tank-top hosing; the clean diffusers are evidently lighter in colour. Bottom: cleaning of polyurethane membranes by means of brushing.

Various methods have been used to clean fine pore diffusers and vary in difficulty and cost. These include cleaning with HCl gas, formic, or acetic acid injected into air distribution lines is also possible (Schmit *et al.*, 1989). Weak organic acids are indicated for membrane diffusers.

Also, pneumatic cleaning is possible for membrane diffusers, labelled flexing and reverse flexing (or relaxing). The practice of flexing consists of periodically increase the air flow to an excess level for a short period of time, which can be followed by the reverse procedure consisting of releasing the air pressure in the air feed and letting the diffuser membrane collapse onto the diffuser housing. While in theory this practice would be effective and relatively simple to implement, seldom is the case when the blowers spare sufficient discharge pressure to meet the effective flexing requirements.

Specific results will depend upon plant design and provisions for diffuser cleaning (Rieth *et al.*, 1990). For example, it is necessary to have spare capacity or periods of reduced loading or modified operation in order to dewater aeration tanks for cleaning. This is generally possible at large plants, but may not be possible at small plants. There are also direct cleaning costs, such as the labour associated with cleaning, chemicals, and replacement parts. Therefore, the choice of cleaning methods and frequencies is non-trivial.

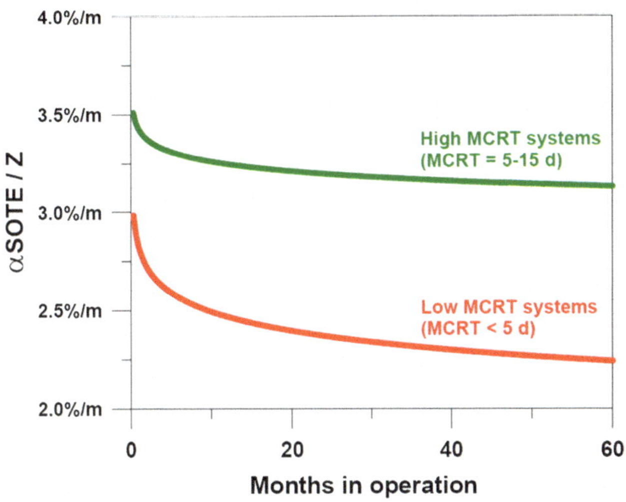

Figure 2.11 Alpha Standard Oxygen Transfer Efficiency per depth (αSOTE/Z) of 103 WWTPs in operation from 0 to 60 months. The WWTPs have been divided into those having a Mean Cell Retention Time (MCRT) less than 5 days and those having a higher MCRT than 5 days.

2.2.5 Mixed liquor concentrations

In the absence of respiration, the physical presence of solids has a detrimental effect on the OTR because the layer of solids that accumulates on the bubbles has low permeability and blocks oxygen transfer (Ju & Sundararajan, 1994). The oxygen transfer has, for a long time, been considered inversely related to the MLSS concentrations in the reactor (Krampe & Krauth, 2003; Muller *et al.*, 1995; Ozdemir & Yenigun, 2013). To represent this phenomenon, different authors have proposed relations between MLSS concentration and the α-factor. For example, Krampe and Krauth (2003) proposed an inverse exponential relation and Henkel *et al.* (2011) theorized an inverse linear relation.

However, Rosso and Stenstrom, (2005) had shown the relationship between the α-factor and MCRT (hence, MLSS) in activated sludge for fine-pore aeration. To conciliate the different positions, we plotted in Figure 2.13 all the data from Rosso and Stenstrom, (2005) and compared it with other data from fine-pore aeration of highly concentrated MLSS reactors. The data sets plotted in Figure 2.13 are from both pilot and full-scale tests. The fitting line is a double exponential curve with descriptive statistics reported in the plot (see Table 2.2 and Figure 2.13). Note the gap between 4 and 6 g/L, where membrane bioreactors are uneconomical to operate and activated sludge clarifiers are limited by the solid gravity flux. In summary, these relationships allow us to obtain similar results for concentrations less than 10 g/L MLSS. However, for higher

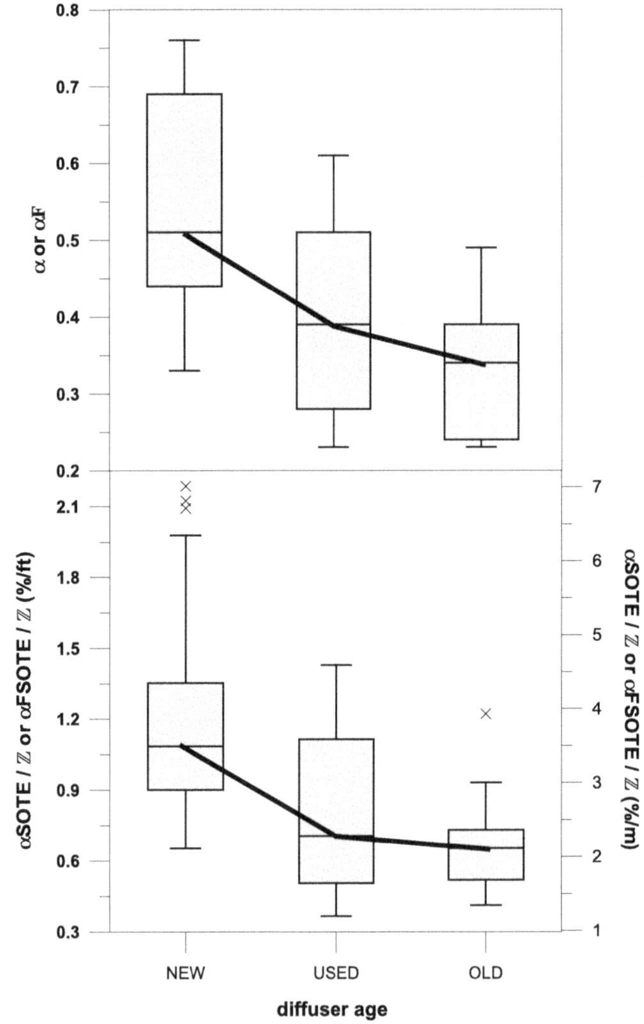

Figure 2.12 Efficiency decrease versus time in operation: $\alpha(F)$ factor on the top, and on the bottom α (F)SOTE/Z. X symbols represent outliers. The black line connects the average values.

concentrations, variations merge because of the type of relationship proposed, and the conditions for validity in each case, as shown in Figure 2.13.

As the sludge concentration increases, the bubble coalescence must increase because of the shear-thinning nature of the sludge. In fact, coalesced bubbles (associated with more interfacial shear) can thin the fluid and experience less resistance to rise. Such coalesce bubbles, having a substantially lower "*a*" in their

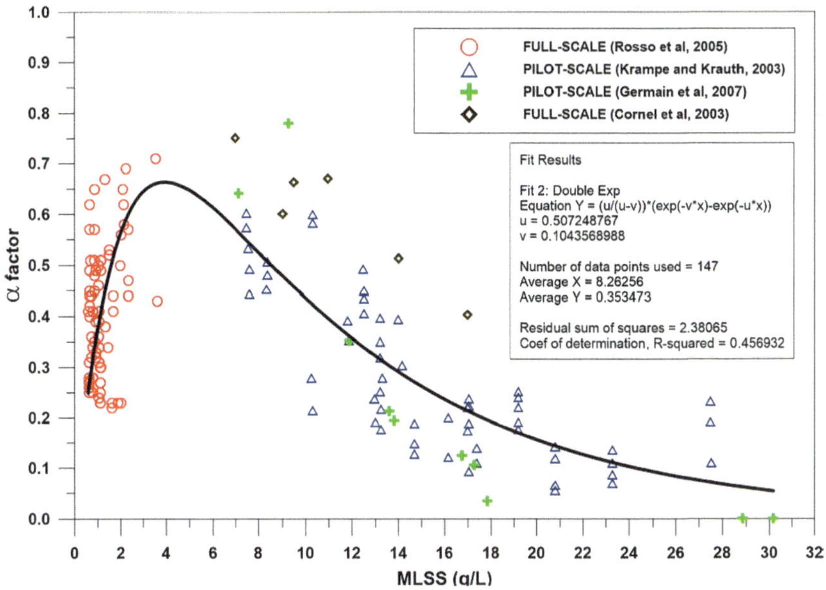

Figure 2.13 Dependence of the α-factor on the MLSS concentration for activated sludge processes (red points) and membrane bioreactors (all other points), equipped with fine-pore diffusers. Note the sharp decrease of alpha due to the heavily non-Newtonian nature of the thick sludge for MLSS >1%.

k_La, do indeed exhibit lower α factors. Correlations between the aeration performance and the apparent viscosity of the activated sludge have been reported. For stirred tank reactors, an exponential correlation exists between the activated sludge k_La and the Reynolds number (Krampe & Krauth, 2003; Nittami *et al.*, 2013). It can be assumed that a rise in apparent viscosity may lead to the production of larger bubbles at the formation stage. Larger bubbles will lead to a reduction of the specific surface area (Durán *et al.*, 2016).

2.2.6 Extracellular polymeric substances

Extracellular polymeric substances (EPS) represent a significant component when discussing microbial aggregates because they hold the aggregates together in a three-dimensional matrix (Sheng *et al.*, 2010). The EPS increase the porosity of the flocs and, therefore, their diffusivity, which is beneficial for oxygen transfer as it is facilitated by the formation of large flocs (Germain *et al.*, 2007).

Germain *et al.* (2007) described a statistical analysis method for identifying the relative effects of various biomass characteristics on oxygen transfer. These authors found that the value of k_La is influenced by the following factors in decreasing order: MLSS > airflow rate > EPS > SMP_{COD} (soluble microbial

Table 2.2 α–factor prediction based on MLSS concentration.

Reference	MLSS Range	Equation
Krampe and Krauth (2003)	0–30 g/L	$\alpha = e^{-0.08788\ \mathrm{MLSS}}$
Henkel *et al.* (2011)	1–12 g/L	$\alpha = -0.062 \times \mathrm{MLVSS} + 0.972 \pm 0.070$
Baquero-Rodríguez *et al.* (2017)	0–30 g/L	$\alpha = \left(\frac{u}{u-v}\right) \cdot \exp(-v \cdot \mathrm{MLSS}) \cdot \exp(-u \cdot \mathrm{MLSS})$ $\{u = 0.507;\ v = 0.104\}$

products). For the α-factor, the influencing factors decreased as follows: MLSS > SMP_{COD} > EPS > airflow rate. Moreover, the authors showed that the k_La and α-factor mirrored aeration, and the EPS (i.e., both increased as aeration and the EPSc increased) were inversely related to the MLSS and SMP_{COD} (the two decreased with increasing levels of MLSS and SMP_{COD}). Overall, the MLSS had the most control over oxygen transfer in the aforementioned investigation, which focused on membrane bioreactors.

2.2.7 The impact of environmental factors

Environmental factors are not under the control of the operator and affect OTE in clean and process water.

2.2.7.1 Temperature

Changes in water temperature induce changes in the oxygen saturation concentration of water. The oxygen saturation concentration of water decreases with increasing temperature, and the decreased solubility of oxygen at higher temperature is an important factor in aeration system design. Increases in the mixed liquor temperature generally coincide with increases in ambient air temperature, and decrease the capacities of the blowers in diffused aeration systems (Jenkins, 2013).

2.2.7.2 Barometric pressure

Despite remarkable progress in aeration, the effects of barometric pressure on OTE were not studied. Regarding aeration energy demand, the blower performance is inversely related to the altitude at which the blower is installed, because of the effects of altitude on atmospheric pressure and air density. Thus, atmospheric pressure has been considered here.

Three factors decrease the air density: increases in air temperature, decreases in atmospheric pressure, and increases in relative humidity. From a mechanical efficiency standpoint, a decrease in air density results in a greater demand for volumetric airflow production by the blower, to guarantee an equivalent airflow mass with respect to standard conditions. The evaluation of energy requirements depends on the operating conditions, control techniques, and blower type (Water Environment Federation, 2009). A comparison of energy consumption at sea level and at higher altitudes is obtained by dividing the energy consumption under standard conditions by the correction factor of the air density under the desired conditions. For cities with elevations above sea level ranging from 1600 to 3600 masl, the corresponding altitude correction factor is in the range between 0.81 and 0.69 (Ludwig, 1997). Such is the case in cities like La Paz, Bolivia (3640 masl), Quito, Ecuador (2850 masl), Bogotá, Colombia (2625 masl), Mexico City, Mexico (2240 masl), and Denver, United States (1600 masl). Most of the above-mentioned locations are considered developing countries with poor sanitation coverage, and since, at the present moment, the development of their WRRFs is largely underway, we highlight here the shortage of published knowledge on aerated processes in high-elevation regions.

2.2.7.3 Saturation concentration of oxygen in water

The oxygen saturation concentration in water is important for aeration and represents the maximum amount of oxygen that can be dissolved in water (Jenkins, 2013). The oxygen saturation concentration in water is a function of barometric pressure, water pressure, salinity, and temperature (Jenkins, 2013), and decreases noticeably when the atmospheric pressure is not 1 atm, and/or the salinity (based on the Practical Salinity Scale) or temperature are high. The procedures for estimating the oxygen saturation concentration under these "standard" conditions can be found in Benson and Krause (1984). As stated before, the driving force for oxygen transfer is the concentration gradient between the oxygen saturation concentration and dissolved oxygen. For all practical purposes, the salinity of potable water can be deemed negligible.

2.2.8 Impact of hydrodynamics

Previous sections of this book summarized the investigations focusing on the impact of environmental and operational conditions of the process on oxygen transfer. There is also an academic interest in exploring the effect on hydrodynamics in gas–liquid transfer driven by both understanding in more depth the transport phenomena within the aeration tank and the desire to model the emission of greenhouse gases.

Although hydraulic design of wastewater treatment systems is a crucial step to assure reliable operation of the process, in the majority of designs flow behaviour is predicted via the ideal reactor model (Stamou, 2008). WRRFs are assumed, then, to have a completely mixed flow regime in all type of tanks (i.e., aerobic, anoxic, anaerobic). However, the hydrodynamics of a bioreactor can influence how "well mixed" is actually an AS system. In fact, in an heterogeneous multi-phase real reactor, the fluid flow can be impacted by vessel geometry, physical properties of its contents (phase, density, viscosity) and operating condition variables (flow rates, distribution of oxygen, temperature, flocs, etc.).

In recent years the approach based on dynamic modelling showed the potential to be a powerful tools providing detailed insight into the unit process and system behaviour. Further details on how to measure the impact of hydrodynamics in WRRFs by CFD and their current limitations can be found in the review by Karpinska and Bridgeman (2016).

At the time this book was written, all the work using computational fluid dynamics to determine the influence of hydrodynamics could not yet be considered in its full state of maturity, as different experts and research groups are still addressing a set of fundamental limitations:

- In some models bubbles are considered part of the liquid phase and are accounted for only as a change in density of the liquid, proportional to the gas holdup fraction; when each phase is considered to have its unique velocity vector, an empirical drag model is employed;
- Bubbles are assumed not to change in size during their rise, despite the change in hydrostatic pressure with water depth, with the consequence that bubbles do not accelerate during their rise and do not increase their perturbation of the liquid velocity field, as in fact happens;
- Bubbles are assumed not to interact to each other;
- Viscosity is the only effect to the gas transfer coefficient, not the effect of contaminants (i.e., surfactants), which is the equivalent to exclusively consider biomass and not the effect of alpha depressing agents (surfactants).

The drive for advancing this knowledge is certainly important, as it will elucidate the gas transfer phenomena whose effects are recorded quantitatively by field-testing (e.g., gas transfer measurements). Therefore, the integration of successful biological wastewater treatment modelling combining hydrodynamics, mass transfer, and biokinetics remains one of the major goals in wastewater process modelling (Le Moullec *et al.*, 2010a, b; Morchain *et al.*, 2014; Pereira *et al.*, 2012).

2.3 TESTING AERATION SYSTEMS

2.3.1 Why we need testing standards

Recalling the section on the definition of roles (Sec. 2.1.1), we need to keep in mind that the main reason for needing testing standards is to establish a transparent level-field for comparison of performance and efficiency, across geographical areas, environmental conditions, type and geometry of equipment, etc. The following sections guide the readers through the American standards, which are equivalent to the European standards in their essence. Small variations between the two may be subject of confusion, hence it is recommended to use consistently the same standard (i.e., American or European) for both clean and process water tests. Mixing the two requires corrections that may lead to errors.

2.3.2 Clean water tests

Clean water testing (ASCE, 2007) can be performed to compare different equipment and configurations. The results are reported as SOTE (%), SOTR ($kg_{O2}\ hr^{-1}$), SAE ($kg_{O2}\ kWh^{-1}$). Care must be exercised when using SAE, since different power measurements can be made. Generally "wire" power is usually preferable, which includes blower, coupling, gearbox and motor inefficiencies. Clean water test results can be used as warranty to verify performance and can also create a competitive bidding environment among manufacturers. Figures 2.14 and 2.15 show a schematic of the experimental setup used in a clean water test and a picture of a clean water test on membrane strip panels before beginning.

2.3.3 Process water tests

For process water conditions, results are reported as OTE, OTR and AE, which include the impacts of non-standard conditions. For off-gas results, it is convenient to use αSOTE, or αSOTR; these two variables are corrected for all non-standard conditions except the α factor. This is possible because the other

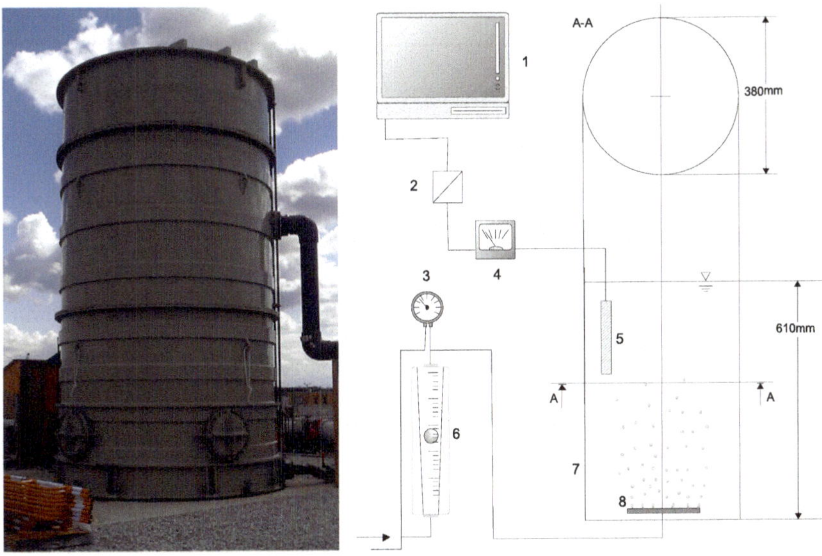

Figure 2.14 Example of clean water test setups. On the left is a full-depth continuous reactor (7.5 m deep; 4.8 m in diameter). On the right is the schematic of a laboratory-scale batch apparatus: (1) data logger; (2) D/A converter; (3) pressure gauge; (4) DO meter; (5) DO probe; (6) air flow meter; (7) aeration tank; (8) aerator.

Figure 2.15 Fine-pore diffusers inside full-depth tanks before clean water testing. On the left are polyurethane strip panels in a 5 m deep tank. On the right are 9 in membrane disc diffusers of various materials in a 7 m tank.

non-standard conditions are easily measured and corrected. The α factor, which is a ratio of mass transfer coefficients in process- to clean-water, can be calculated from off-gas results if clean water data are available (i.e., *a posteriori*). In fouled aeration systems, a second parameter, F, is used to define the degree of fouling. Therefore the efficiency of a new fine pore aeration system could be defined by αSOTE and a used or fouled system by αFSOTE. αSOTE or αFSOTE are used for process water transfer efficiencies. To compare the results presented here to actual process conditions, the other corrections, such as DO concentration and temperature must be applied (see Sec. 2.1).

In order to better define aerator performance, the off-gas testing technique has been extensively used to measure diffused aeration efficiency (Fig. 2.16). Off-gas testing was developed by Redmon *et al.* (1983) in conjunction with the US EPA sponsored ASCE Oxygen Transfer Standards Committee. This committee produced a fine pore manual (US EPA, 1989), a clean water oxygen transfer standard (ASCE, 1984, 1991, 2007) and guidelines for process water

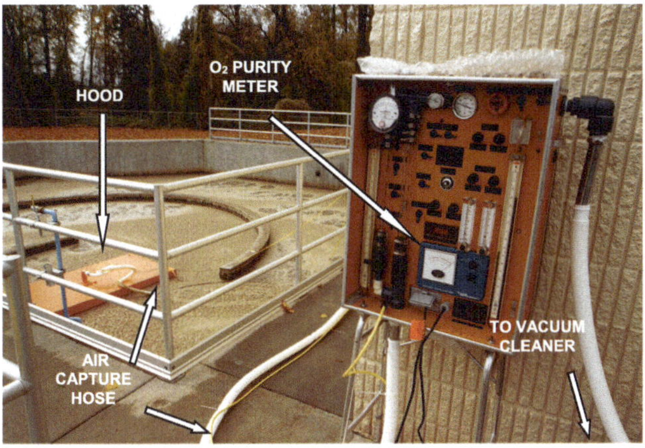

Figure 2.16 Off-gas testing setup.

testing (Warriner, 1996). Clean water testing and off-gas testing are described in detail in these publications.

Off-gas is the gas emitted from the surface of the liquid volume being aerated. By measuring the oxygen concentration in off-gas, and determining the oxygen concentrations or molar percentages (e.g., 20.9% for air, 100% for pure oxygen) of the gas being diffused into the system, the OTE can be calculated from any location within the aerobic reactor. Ample evidence is available that demonstrates how off-gas testing is the best technique for determining the *in situ* performance of diffused aeration systems. The basic principles of off-gas testing involve studying the oxygen mass balance of the volume or system being aerated.

2.3.3.1 The evolution of the off-gas

The first reference to a technique similar to what is currently referred to as off-gas analysis for water resource recovery was presented by Sawyer and Nichols (1939). When this technique was first used, the equipment was called an "Oxy-Utilometer," and a volumetric method was used to measure the amount of oxygen consumed by the activated sludge during water resource recovery. Decades later, starting in the 1960s, discussions of the OTE in aeration systems for water resource recovery that used techniques similar to off-gas analysis appeared in the scientific literature. These analyses were conducted to define aeration system performance, with each analysis opting for a different focus, including considering the volume of air used per mass of biological oxygen demand removed, determining the diffuser placement pattern that would provide the best oxygen transfer performance, or comparing different diffuser configurations (Leary *et al.*, 1968). In the 1980s, Boyle and Redmon (1983) presented generalities for the functions of off-gas analysers (as they are known today), and guidelines for their use in field studies. In the 1980s, off-gas analysers were composed of the following four principal components: (1) a floating hood to capture the gas; (2) a hose connecting the hood to the analytical circuit; (3) an analytical circuit for monitoring the off-gas composition, temperature, pressure, and gas flow rate; and (4) a vacuum source for drawing gas from the hood through the analytical circuit.

Regarding recent trends connected to real time off-gas monitoring, the work performed by Libra *et al.* (2002) is notable. These authors reported experiences and results related to the construction and operation of off-gas analysers coupled with floating collection hoods used in 24 h experiments. An evolution of these was by Leu *et al.* (2009), who directly measured the OTE and airflow rate through a flow pipe that collects data at hourly intervals using semi-automatic analyzers. In their work, data from WRRFs in different southern California were gathered. A long-term demonstration of fully aturomated analyzers employed in four plants over periods spanning 3 to 12 months was presented by Jeung *et al.* (2013). The results of their study indicated that real-time off-gas monitoring can be applied to continuously monitor the efficiencies of wastewater aeration processes over extended periods.

A configuration of the off-gas analyser, similar to that described by Redmon *et al.* (1983) is presented in the ASCE 18-96 standard (American Society of Civil Engineers, 1997). This document also contains a graphic representation of the analyser that is still employed today. Details regarding the required materials (including costs), drawbacks, instrument and hood assembly for permanent installation, operation and maintenance of off-gas analysers, data processing information, and reference analysis are available in Jeung *et al.* (2013).

2.3.3.2 Off-gas advantages

The net result of the improved testing methods is an increase in accuracy and precision in designing and quantifying aeration systems. These methods are now widely used in the United States (e.g., Iranpour & Stenstrom, 2001; Libra *et al.*, 2002; Groves *et al.*, 1992), Europe (e.g., Gillot *et al.*, 2005). Off-gas analysis is also being proposed as an additional aeration control mechanism (Trillo *et al.*, 2004).

Overall, off-gas monitoring can be used to determine the best options for designing and expanding aeration systems. Examples of analyses that can be conducted with off-gas monitoring while considering cost-benefit ratios include the following: (1) evaluation of several diffuser types in side-by-side tests under process conditions; (2) evaluation of diffuser fouling problems and diffuser cleaning procedures in terms of effectiveness; (3) optimization of cleaning schedules by comparing the energy required and cleaning costs; (4) comparison of fixed and variable flow blowers, and the evaluation of the benefits of using an equalization basin; (5) evaluation of aeration system control procedures; and

(6) real-time analysis of OTE signals which can have dynamic operational implications, such as feed-forward off-gas control and energy minimization when connected to facility data systems (Boyle & Redmon, 1983; Jeung *et al.*, 2013; Trillo *et al.*, 2004).

2.3.4 Testing pitfalls

2.3.4.1 Dissolved oxygen probe response time

As early as 1974, Boyle *et al.* had recognised the pitfalls in quantifying oxygen transfer parameters. Philichi and Stenstrom (1989) outlined a procedure for establishing the response time of a sensor and its effect on estimating k_La. The typical response time of polarographic sensors with highly sensitive membranes is approximately 8 seconds (Table 2.3), and for optic sensors the response time approaches 40 seconds. The development of standard procedures for estimating oxygen transfer parameters began in the 1980s (Brenner, 1983). At that time, optical dissolved oxygen probes were not commercially available. The first references discussing the applications of optical dissolved oxygen probes for water resource recovery were published in 2000 (Santos & Farahi, 2014). This technology should be considered for oxygen transfer parameter estimation tests because of its recent development and its wide commercial availability. According to the literature review, the effects of measuring response time on the optical sensors, when they are applied for estimating the k_La, have not yet been considered. Nevertheless, recent findings contradict previous studies which claim that lag time in polarographic dissolved oxygen probes response can bias estimates of k_La. On the basis of the results obtained for the k_La, estimated using different types of dissolved oxygen probes (optic and polarographic) in a small tank, Baquero-Rodríguez and Lara-Borrero (2016) concluded that there could be

Table 2.3 Time response from different commercially available Dissolved Oxygen (DO) probes and their effect on measuring the oxygen transfer rate (K_La).

DO Probe Type	DO Probe t_c	Max K_La Measurable	
	s	h^{-1}	s^{-1}
Electrochemical	1	72.0	0.0200
Electrochemical	2	36.0	0.0100
Electrochemical	6	12.0	0.0033
Electrochemical	5	14.4	0.0040
Electrochemical	10	7.2	0.0020
Optical	10	7.2	0.0020
Optical	30	2.4	0.0007
Optical	60	1.2	0.0003

no significant influence of the probe lag (i.e., time constant) on the parameters used to assess oxygen transfer efficiency in clean water for k_La as high as 20 h^{-1}. However, care must be used when extending such results to full-scale tanks, therefore electrochemical sensors are still to be considered the technology of reference for full-scale testing.

Each type of dissolved oxygen sensor (optic and polarographic) has advantages and disadvantages. From a WWTP/WRRF operator viewpoint, optical sensors require less maintenance than polarographic sensors and have less noise due to their slower response. From a practical viewpoint, lower maintenance and stable readings are attractive features when selecting dissolved oxygen sensors. Additional details regarding optical sensor specifications are available by their corresponding manufacturers.

2.4 REFERENCES

Alkhalidi A. A. T. and Amano R. S. (2015). Factors affecting fine bubble creation and bubble size for activated sludge. *Water Environment Journal*, **29**, 105–113.

Amaral A., Schraa O., Rieger L., Gillot S., Fayolle Y., Bellandi G., Amerlinck Y., Mortier S. T. F. C., Gori R., Neves R. and Nopens I. (2016). Towards advanced aeration modelling: from blower to bubbles to bulk. *Water Science and Technology*, **75**, 507–517.

Amerlinck Y., Bellandi G., Amaral A., Weijers S. and Nopens I. (2016). Detailed off-gas measurements for improved modelling of the aeration performance at the WWTP of Eindhoven. *Water Science and Technology*, **74**, 203–211.

Andersen H., Siegrist H., Bent H.-S. and Ternes A. T. (2003). Fate of estrogens in a municipal sewage treatment plant. *Environmental Science and Technology*, **37**, 4021–4026.

ASCE (1984). *Standard for the Measurement of Oxygen Transfer in Clean Water*. American Society of Civil Engineers, Reston, VA, USA, pp. 46.

ASCE (1991). *ASCE Standard: Measurement of Oxygen Transfer in Clean Water*. American Society of Civil Engineers, Reston, VA, USA.

ASCE (1997). *ASCE Standard: Standard Guidelines for In-Process Oxygen Transfer Testing (18–96)*. New York.

ASCE (2007). *Measurement of Oxygen Transfer in Clean Water, ASCE Standard ASCE/EWRI 2-06*. American Society of Civil Engineers, Reston, VA, USA.

Baquero-Rodríguez G. A. and Lara-Borrero J. A. (2016). The influence of optic and polarographic dissolved oxygen sensors estimating the volumetric oxygen mass transfer coefficient (K_La). *Modern Applied Science*, **10**, 142–151.

Baquero-Rodríguez G. A., Lara-Borrero J. A., Nolasco D. and Rosso D. (2017). A critical review of the factors affecting modeling oxygen transfer by fine-pore diffusers in activated sludge. *Water Environment Research*, **90**, 431–441.

Benson B. B. and Krause D. Jr. (1984). The concentration and isotopic fractionation of oxygen dissolved in freshwater and seawater in equilibrium with the atmosphere. *Limnology and Oceanography*, **29**, 620–32.

Bouaifi M. and Roustan M. (2001). Power consumption, mixing time and homogenisation energy in dual-impeller agitated gas–liquid reactors. *Chemical Engineering and Processing: Process Intensification*, **40**, 87–95.

Bouaifi M., Hebrard G., Bastoul D. and Roustan M. (2001). A comparative study of gas hold-up, bubble size, interfacial area and mass transfer coefficients in stirred gas–liquid reactors and bubble columns. *Chemical Engineering and Processing: Process Intensification*, **40**, 97–111.
Boyle W. C. and Redmon D. T. (1983). Biological fouling of fine bubble diffusers: state-of-art. *Journal of Environmental Engineering*, **109**, 991–1005.
Boyle W. C., Berthouex P. M. and Rooney T. C. (1974). Pitfalls in parameter estimation for oxygen transfer data. *Journal of Environmental Engineering*, **100**, 391–408.
Brenner R. C. (1983). *Development of Standard Procedures for Evaluating Oxygen Transfer Devices (EPA-600/2-83-102)*. US EPA, Cincinnati, OH, USA.
Clift R., Grace J. and Weber M. E. (2005). *Bubbles, Drops, and Particles*. Dover, Mineola, New York, USA.
Cornel P., Wagner M. and Krause S. (2003). Investigation of oxygen transfer rates in full scale membrane bioreactors. *Water Science and Technology*, **47**(11), 313–319.
Downing A. and Boon A. G. (1963) Oxygen transfer in the activated sludge process. In: *Advances in Biological Wastewater Treatment*, W. W. Eckenfelder and J. McCabe (eds), Macmillan, New York, NY, USA, pp. 131–147.
Durán C., Fayolle Y., Pechaud Y., Cockx A. and Gillot S. (2016). Impact of suspended solids on the activated sludge non-newtonian behaviour and on oxygen transfer in a bubble column. *Chemical Engineering Science*, **141**, 154–165.
Eckenfelder W. W. (1959). Factors affecting the aeration efficiency of sewage and industrial wastes. *Sewage Industrial Wastes*, **31**, 60–70.
Eckenfelder W. W. and Barnhart E. L. (1961). The effect of organic substances on the transfer of oxygen from air bubbles in water. *AIChE Journal*, **7**, 631–634.
Eckenfelder W. and Ford D. (1968) Engineering aspects of surface aeration design. In: *Proceedings of 22nd Industrial Waste Conference*. Purdue University, Lafayette, Indiana, USA.
Fayolle Y., Cockx A., Gillot S., Roustan M. and Héduit A. (2007). Oxygen transfer prediction in aeration tanks using CFD. *Chemical Engineering Science*, **62**, 7163–7171.
Fayolle Y., Gillot S., Cockx A., Bensimhon L., Roustan M. and Héduit A. (2010). *In situ* characterization of local hydrodynamic parameters in closed-loop aeration tanks. *Chemical Engineering Journal*, **158**, 207–212.
Germain E., Nelles F., Drews A., Pearce P., Kraume M., Reid E., Judd S. J. and Stephenson T. (2007). Biomass effects on oxygen transfer in membrane bioreactors. *Water Research*, **41**, 1038–1044.
Gillot S. and Héduit A. (2008). Prediction of alpha factor values for fine pore aeration systems. *Water Science and Technology*, **57**, 1265–1269.
Gillot S., Capela-Marsal S., Roustan M. and Héduit A. (2005). Predicting oxygen transfer of fine bubble diffused aeration systems – model issued from dimensional analysis. *Water Research*, **39**, 1379–1387.
Göbel A., McArdell C. S., Joss A., Siegrist H. and Giger W. (2007). Fate of sulfonamides, macrolides, and trimethoprim in different wastewater treatment technologies. *Science of the Total Environment*, **372**, 361–371.
Groves K. P., Daigger G., Simpkin T., Redmon D. T. and Ewing L. (1992). Evaluation of oxygen transfer efficiency and alpha-factor on a variety of diffused aeration systems. *Water Environment Research*, **64**, 691–698.

Harper W. F. Jr. and Jenkins D. (2003). The effect of an initial anaerobic zone on the nutrient requirements of activated sludge. *Water Environment Research*, **75**, 216–224.

Henkel J. (2010) Oxygen transfer phenomena in activated sludge. PhD Thesis, TU Darmstadt, Darmstadt, Germany.

Henkel J., Cornel P. and Wagner M. (2011). Oxygen transfer in activated sludge – new insights and potentials for cost saving. *Water Science and Technology*, **63**, 3034–3038.

Hwang H. J. and Stenstrom M. K. (1979). *The Effects of Surface Active Agents on Oxygen Transfer*. Water Resources Program, School of Engineering and Applied Science, University of Califonia.

Iranpour R. and Stenstrom M. K. (2001). Relationship between oxygen transfer rate and airflow for fine-pore aeration under process conditions. *Water Environment Research*, **73**, 266–275.

Jenkins T. E. (2013). *Aeration Control System Design : a Practical Guide to Energy and Process Optimization*. Wiley, Chichester, UK.

Jeung M. K., Larson L. E., Stenstrom M. K. and Rosso D. (2013). Dynamic energy footprint monitoring of wastewater aeration systems via full-scale deployment of real-time off-gas analyzers. *Proceedings of the Water Environment Federation*, **2013**, 291–299.

Ji L. and Zhou J. (2006). Influence of aeration on microbial polymers and membrane fouling in submerged membrane bioreactors. *Journal of Membrane Science*, **276**, 168–177.

Jiang L.-M., Garrido-Baserba M., Nolasco D., Al-Omari A., DeClippeleir H., Murthy S. and Rosso D. (2017). Modelling oxygen transfer using dynamic alpha factors. *Water Research*, **124**, 139–148.

Ju L.-K. and Sundararajan A. (1994). The effects of cells on oxygen transfer in bioreactors: physical presence of cells as solid particles. *Chemical Engineering Journal Biochemical Engineering Journal*, **56**, B15–B21.

Kaliman A., Rosso D., Leu S. S. Y. and Stenstrom M. K. (2008). Fine-pore aeration diffusers: accelerated membrane ageing studies. *Water Research*, **42**, 467–475.

Karpinska A. M. and Bridgeman J. (2016). CFD-aided modelling of activated sludge systems – a critical review. *Water Research*, **88**, 861–879.

Kessener H. and Ribbius F. (1934). Comparison of aeration systems for the activated sludge process. *Sewage Works Journal*, **6**, 423–443.

Krampe J. and Krauth K. (2003). Oxygen transfer into activated sludge with high MLSS concentrations. *Water Science and Technology*, **47**(11), 297–303.

Leary R. D., Ernest L. A. and Katz W. J. (1968). Effect of oxygen-transfer capabilities on wastewater treatment plant performance. *Journal of the Water Pollution Control Federation*, **40**, 1298–1310.

Le Moullec Y., Gentric C., Potier O. and Leclerc J. P. (2010a). CFD simulation of the hydrodynamics and reactions in an activated sludge channel reactor of wastewater treatment. *Chemical Engineering Science*, **65**, 492–498.

Le Moullec Y., Gentric C., Potier O. and Leclerc J. P. (2010b). Comparison of systemic, compartmental and CFD modelling approaches: application to the simulation of a biological reactor of wastewater treatment. *Chemical Engineering Science*, **65**, 343–350.

Leu S.-Y., Rosso D., Larson L. E. and Stenstrom M. K. (2009). Real-time aeration efficiency monitoring in the activated sludge process and methods to reduce energy consumption and operating costs. *Water Environment Research*, **81**, 2471–2481.

Lewis W. K. and Whitman W. G. (1924). Principles of gas absorption. *Industrial and Engineering Chemistry*, **16**, 1215–1220.

Libra J. A., Schuchardt A., Sahlmann C., Handschag J., Wiesmann U. and Gnirss R. (2002). Comparison of the efficiency of large-scale ceramic and membrane aeration systems with the dynamic off-gas method. *Water Science and Technology*, **46**(4–5), 317–324.

Loubière K. and Hébrard G. (2004). Influence of liquid surface tension (surfactants) on bubble formation at rigid and flexible orifices. *Chemical Engineering and Processing: Process Intensification*, **43**, 1361–1369.

Ludwig E. E. (1997). *Applied Process Design for Chemical and Petrochemical Plants*. Gulf Professional Publishing, Houston, TX, USA.

Mancy K. H. and Okun D. A. (1960). Automatic recording of dissolved oxygen in aqueous systems containing surface active agents. *Analytical Chemistry*, **32**, 108–114.

Martins A. M. P., Heijnen J. J. and van Loosdrecht M. C. M. (2004). Bulking sludge in biological nutrient removal systems. *Biotechnology and Bioengineering*, **86**, 125–135.

McGinnis D. F. and Little J. C. (2002). Predicting diffused-bubble oxygen transfer rate using the discrete-bubble model. *Water Research*, **36**, 4627–4635.

Metcalf and Eddy 2014. *Wastewater Engineering: Treatment and Resource Recovery*, 5th Edn. McGraw-Hill Education, New York, USA.

Morchain J., Gabelle J.-C. and Cockx A. (2014). A coupled population balance model and CFD approach for the simulation of mixing issues in lab-scale and industrial bioreactors. *AIChE J*, **60**, 27–40.

Muller E. B., Stouthamer A. H., van Verseveld H. W. and Eikelboom D. H. (1995). Aerobic domestic waste water treatment in a pilot plant with complete sludge retention by cross-flow filtration. *Water Research*, **29**, 1179–1189.

Nittami T., Katoh T. and Matsumoto K. (2013). Modification of oxygen transfer rates in activated sludge with its characteristic changes by the addition of organic polyelectrolyte. *Chemical Engineering Journal*, **225**, 673–678.

Ozdemir B. and Yenigun O. (2013). A pilot scale study on high biomass systems: energy and cost analysis of sludge production. *Journal of Membrane Science*, **428**, 589–597.

Parker D., Geary S., Jones G., McIntyre L., Oppenheim S., Pedregon V., Pope R., Richards T., Voigt C., Volpe G., Willis J. and Witzgall R. (2003). Making classifying selectors work for foam elimination in the activated-sludge process. *Water Environment Research*, **75**, 83–91.

Pereira J. P., Karpinska A., Gomes P. J., Martins A. A., Dias M. M., Lopes J. C. B. and Santos R. J. (2012) Activated sludge models coupled to CFD simulations. In: *Single and Two-Phase Flows on Chemical and Biomedical Engineering*, R. Dias, R. Lima, A. A. Martins and T. M. Mata (eds). Bentham Science, Sharjah, UAE, pp. 153–173.

Philichi T. and Stenstrom M. K. (1989). The effect of dissolved oxygen probe lag upon oxygen transfer parameter estimation. *Journal of the Water Pollution Control Federation*, **61**, 83–86.

Redmon D., Boyle W. C. and Ewing L. (1983). Oxygen transfer efficiency measurements in mixed liquor using off-gas techniques. *Journal of the Water Pollution Control Federation*, **55**, 1338–1347.

Rieth M., Boyle W. and Ewing L. (1990). Effects of selected design parameters on the fouling characteristics of ceramic diffusers. *Journal of the Water Pollution Control Federation*, **62**, 877–886.

Rosso D. and Al-Omari A. (2018). Carbon capture and management strategies for energy harvest from wastewater. *Water Environment Research Foundation* (Project U3R14). [online] In press.

Rosso D. and Stenstrom M. K. (2005). Comparative economic analysis of the impacts of mean cell retention time and denitrification on aeration systems. *Water Research*, **39**, 3773–3780.

Rosso D. and Stenstrom M. K. (2006a). Economic implications of fine-pore diffuser aging. *Water Environment Research*, **78**, 810–815.

Rosso D. and Stenstrom M. K. (2006b). Surfactant effects on α-factors in aeration systems. *Water Research*, **40**, 1397–1404.

Rosso D. and Stenstrom M. K. (2007). Energy-saving benefits of denitrification. *Environmental Engineer: Applied Research and Practice*, **2007**, 29–38.

Rosso D., Jiang L. M., Hayden D. M., Pitt P., Hocking C. S., Murthy S. and Stenstrom M. K. (2012). Towards more accurate design and specification of aeration systems using on-site column testing. *Water Science and Technology*, **66**, 627–634.

Roy S. and Duke S. R. (2004). Visualization of oxygen concentration fields and measurement of concentration gradients at bubble surfaces in surfactant-contaminated water. *Experiments in Fluids*, **36**, 654–662.

Santos J. and Farahi F. (2014). *Handbook of Optical Sensors*. CRC Press, Boca Raton, FL, USA.

Sawyer C. N. and Nichols M. S. (1939). Activated sludge oxidations: I effect of sludge concentration and temperature upon oxygen utilization. *Sewage Works Journal*, 51–67.

Schmit F. L., Redmon D. T. and Ewing L. (1989) US Patent 4889620A, In place gas cleaning of diffusion elements.

Schuler A. J. and Jang H. (2007). Density effects on activated sludge zone settling velocities. *Water Research*, **41**, 1814–1822.

Sheng G.-P., Yu H.-Q. and Li X.-Y. (2010). Extracellular polymeric substances (EPS) of microbial aggregates in biological wastewater treatment systems: a review. *Biotechnological Advances*, **28**, 882–894.

Soliman M. A., Pedersen J. A., Park H., Castaneda-Jimenez A., Stenstrom M. K. and Suffet I. H. (Mel) (2007). Human pharmaceuticals, antioxidants, and plasticizers in wastewater treatment plant and water reclamation plant effluents. *Water Environment Research*, **79**, 156–167.

Stamou A. I. (2008). Improving the hydraulic efficiency of water process tanks using CFD models. *Chemical Engineering and Processing: Process Intensification*, **47**, 1179–1189.

Stenstrom M. K. and Gilbert R. G. (1981). Effects of alpha, beta and theta factor upon the design, specification and operation of aeration systems. *Water Research*, **15**, 643–654.

Tampus M. V., Martins A. M. P. and van Loosdrecht M. C. M. (2004). The effect of anoxic selectors on sludge bulking. *Water Science and Technology*, **50**(6), 261–268.

Trillo I., Jenkins T. E., Redmon D., Hilgart T. and Trillo J. (2004). Implementation of feedforward aeration control using on-line offgas analysis: the Grafton WWTP experience. *Proceedings of the Water Environmation Federation*, **2004**, 27–45.

Tseng L. Y., Gori R. and Rosso D. (2015). Effects of activated sludge process conditions on the production of extracellular polymeric substances: results of yearlong monitoring in a warm climate. *Environmental Engineering Science*, **32**(7), 582–592.

US EPA (1989). *Design Manual: Fine Pore Aeration Systems.* (EPA/625/1-89/023) US Environmental Protection Agency, Washington, DC, USA.
Vogelaar J. C. T., Klapwijk A., van Lier J. B. and Rulkens W. H. (2000). Temperature effects on the oxygen transfer rate between 20 and 55°C. *Water Research*, **34**, 1037–1041.
Wagner M. R. and Pöpel H. J. (1998). Oxygen transfer and aeration efficiency – influence of diffuser submergence, diffuser density, and blower type. *Water Science and Technology*, **38**(3), 1–6.
Warriner R. (1996). *Oxygen Transfer Efficiency Surveys at the South Shore Wastewater Treatment Plant 1985–1987 (EPA/600/R-94/099).* US Environmental Protection Agency, Cincinnati, OH, USA.
Water Environment Federation (2009). *Energy Conservation in Water and Wastewater Facilities – MOP 32*, 1st edn. McGraw-Hill Professional, New York, USA.

Chapter 3

Mixing in activated sludge systems

Coenraad Pretorius[1], Manel Garrido-Baserba[2] and Diego Rosso[2]

[1]CDM Smith

[2]University of California, Irvine

3.1 INTRODUCTION

Aeration and mixing are key to maintaining a healthy, stabilized biological population and thus achieving optimal treatment. In wastewater treatment plants, aeration typically provides both oxygen and mixing. However, due to plant hydrodynamics, process conditions in aeration basins can vary significantly along the tank length, depth and, sometimes, width. Strong gradients for DO, nutrients, food to mass ratios and floc sizes (and types) can be found limiting the effective reactor volume where both physical and biological conversions are taking place. The uniformity, or homogeneous characteristics of the activated sludge suspension, at both macroenvironment (bioreactor) and microenvironment (floc) scales is determined by the mixing conditions. Short-circuiting can make up as much as 50–80% of the inflow and reducing conversion efficiencies (Ahnert *et al.*, 2010), while the corresponding dead zones can amount to 10–20% of basin volume (De Clercq *et al.*, 1999).

Mixing is essential to accomplish the following objectives:

(1) To keep mixed liquor in suspension and ensure that the entire zone volume is utilized

(2) To blend different feed streams to produce a homogeneous suspension

Editor: Diego Rosso
doi: 10.2166/9781780407845_0073

(3) To encourage flocculation of the mixed liquor
(4) To avoid short-circuiting.

Generally, mixing can be classified in two types according the length of the process. The first is rapid (less than 30s) and used where blending is required, such as where a chemical is added to the effluent or mixed liquor (i.e., addition of salts or alum for flocculation, polymers for dewatering, hypochlorite for disinfection, etc.) or different mixed liquor streams are combined (i.e., RAS and PE, internal recycle streams and mainstream flow). This type of rapid mixing has been studied and modelled extensively. The objective is relatively straightforward: ensure that the chemical is well dispersed in the effluent. Details, equipment characteristics, and further examples used in wastewater treatment facilities for rapid mixing can be found in references such as Metcalf and Eddy (2014).

On the other hand, the mixing that is less understood and certainly more difficult to model or to monitor is the continuous (or gentle) mixing, which is strictly intertwined with microbial reaction rates and governs two key processes to accelerate the biological conversion process: flocculation (formation of microbial aggregates or flocs) and maintenance of solids in suspension. Both are necessary to homogenize the content of the reactor (or digester). Some texts seek to distinguish between these two goals, but it will be argued here that they are similar in most applications, given that solids need to be maintained in suspension in order for flocculation to occur. Understanding and controlling mixing, especially related to aeration and microbial conversion rates, have been largely overlooked, limiting its potential. This chapter will discuss mixing in both aerated and unaerated zones, but with special focus on the required mixing conditions for biological processes to take place, while less attention will be paid to applications as blending of chemicals (i.e., flocculants, coagulants, chlorine, etc.) or miscible liquids.

The trend toward excessive power use in wastewater treatment associated with a (sometimes overly) conservative design approach has in recent years been countered by the increasing awareness that it is in fact possible to achieve energy neutrality (and even become a net power producer) while treating wastewater. Given that mixing and aerating in activated sludge tanks are energy-intensive processes, it has become important to optimize mixing equipment (Füreder *et al.*, 2018; Sharma *et al.*, 2011). While aeration can make up about 50–75% of the total energy consumption of a WRRF some authors has shown that mixing consumes about 5–20% (Füreder *et al.*, 2018; Krampe, 2011) and up to 75% of the tanks have an optimization potential regarding energy consumption (Füreder *et al.*, 2018). Perhaps the first step towards energy neutrality or net power production should be to ensure that mixing and aeration power are properly optimized, so that waste of power is eliminated.

This chapter reviews methods for estimating the power required for mixing, approaches for quantifying the degree of mixing, methods for sizing and designing mixers, available mixer technologies as well as attempts to optimize mixer power use, and discusses conclusions and recommendations.

3.1.1 Mixing and aeration: either, both or neither?

Aeration has been always intrinsically associated with mixing, whether surface aeration or diffused air is used. The release of a continuous supply of bubbles satisfies the oxygen demand, but also maintains the mixing conditions within the reactor, which helps to ensure that the corresponding substrate (i.e., COD, NO_3^-, VFAs, etc.) is in contact with microorganisms. Diffused air is, for example, used to fulfil both the mixing and oxygen requirements. However, there are a few scenarios, very common in current WRRFs seeking improved nutrient removal and process optimization, where decoupling aeration from mixing may deliver substantial advantages. Avoiding aeration while maintaining the advantages related to mixing is sometimes a priority. While avoiding aeration improves the removal of nutrients by enhancing anoxic and anaerobic conditions (i.e., denitrification, PAO enhancement, etc.), keeping mixing conditions is key in order to: (1) maintain channel velocities; (2) keep the reactor content mixed and avoiding settling; and (3) aid the destratification and always maintain the required anoxic or anaerobic conditions. In unaerated zones, on the other hand, mixing must be provided with minimal aeration.

Providing only mixing without aeration can be done by two different approaches. The most obvious is by using mechanical means (see Sec. 3.4.1), impellers being the most common alternative. However, there is also a far less obvious approach, which consists in using air-powered equipment to provide mixing, but not aeration (see Sec. 3.4.3.1). Air-powered mixers have the potential to support denitrification and even phosphorus release at lower power level (W/m^3) than mechanical mixers, provided that oxygen transfer can be minimized.

Swing zones, that is zones designed to be operated in either aerated or unaerated mode, present several challenges for designers: primarily how to provide both mixing and aeration equipment cost-effectively, while preventing the operation of one from damaging the other. One popular approach is to provide mechanical mixers and membrane fine-bubble diffusers. The designer must ensure that operating the mixer would not result in local fluid velocities high enough to dislodge any part of the aeration system. Another concern is that idling the diffusers increases the rate of fouling significantly, so that operators may find diffusers incapable of releasing bubbles after being out of service for an extended period of time.

One way to address this concern would be to install and operate a single piece of equipment that can provide both mixing and aeration. Jet aerators are one option where the air supply can be turned off while liquid flow can be maintained for mixing. Jet mixer efficiency is discussed later in this chapter. Another option is the hyperbolic mixer/aerator. The mixing efficiency of hyperboloid mixers has been studied extensively, but the aeration efficiency not as much. A final option would be to provide a dual dropleg fine-bubble aeration system, where a one dropleg is used for aeration and a smaller dropleg supplying air to strategically placed diffusers is used for mixing only. This option would not address the concern about diffusers experiencing higher rates to fouling when not in service.

Thus, de-coupling aeration from aeration needs to be understood and adequately implemented in order to maximize some specific biological processes. Other scenarios can be found where decoupling both concepts can provide significant benefits for the operation of the treatment facility. For example, sometimes limiting or reducing both aeration and mixing conditions can provide improved nutrient removal. A relatively recent development (Barnard *et al.*, 2017) is to provide incomplete mixing in unaerated zones to induce *in situ* fermentation. Incomplete mixing conditions aim to allow solids blankets developing at the floor of the unaerated zones, triggering anaerobic conditions in the blanket and promoting fermentation-related processes, which transform complex organic carbon into more easily degradable and available organic carbon.

3.1.2 Conservative approach: a typical headache

It has been realized that the conservative approach of oversizing mixing systems during the design phase can have major disadvantages (Pretorius *et al.*, 2015). In this section, one of the main problems facing treatment facilities will be addressed in order to highlight the importance of designing properly sized mixing equipment.

The following consequences can appear when a conservative designing approach is adopted:

- excessive capital is spent on equipment and too much power is used for the entire service life of the mixer, wasting resources and negatively impacting the environment;
- over-mixing in anaerobic or anoxic zones can lead to entrainment of air, negatively impacting selector performance and effluent quality, especially in biological nutrient removal (BNR) systems;
- the mixed-liquor floc structure can be damaged, encouraging floc breakup and negatively impacting sludge settleability.

Using an overly conservative mixing criterion means wasting power, over-aerating (leading to high DO concentrations which may in turn interfere with the operation of unaerated zones receiving flow from the over aerated zone, leading to reduced biological nutrient removal and/or additional operating cost if external substrates are used to restore nutrient removal) and, potentially, de-flocculation of mixed liquor negatively affecting sludge settling characteristics and even reducing secondary treatment capacity.

Similarly, Barnard *et al.* (2017) stated that overly conservatively sized mixers often inhibit the fermentation of readily biodegradable COD in anaerobic zones, owing to incidental aeration, thus limiting the degree of biological phosphorus removal that can be achieved. Furthermore, it is possible that larger flocs would allow lower oxidation–reduction potential (ORP) values at the core of the floc, as oxygen (and potentially nitrate) is consumed before it can diffuse to the core. Low ORP at the inner layer maximizes denitrifying conditions, and phosphorus release. This means that improved nutrient removal due to increased floc size could be obtained by avoiding excessive mixing conditions.

Taking into consideration the previous disadvantages, the importance to properly design mixing equipment is highlighted. Therefore, attention needs to be paid to those scenarios where conservative design approaches were (or are) implemented, as the use of the term *conservative* is often misused and implies equipment oversizing characteristics, which will result in wasting of money, degradation of the effluent quality and limiting the plant capacity.

3.1.3 Mixing and flocculation

New approaches are expected to provide further understanding of the existing limitations concerning the link between mixing and the rates at which contaminants are removed biologically (or conversion rates). New approaches (from DO and OUR mapping to ORP and temperature gradient mapping) are emerging to help improve our understanding of the relationship between mixing and biological activity, aiming to maximize and unveil the hidden potential for process optimization.

It has been found that increasing mixing (i.e., impeller speeds above the level required for maintaining sludge suspension) does not increase the rate at which organic matter and nutrients are converted to biogenic CO_2, water and beneficial bacteria. Pollutants transformation is limited by the mass transfer between the flocs (i.e., suspended bacterial aggregates) and the surrounding fluid. To increase removal rates by promoting an increase in mass transfer the so-called slip velocity, defined as the relative velocity between solids (i.e., flocs) and fluid (containing soluble constituents), would need to be increased. However, in aeration basins where the suspended bacterial flocs are in suspension, increasing the fluid velocity will not correspond to an increase in slip velocity (Paul *et al.*, 2004). Once a sufficient level of homogeneity has been achieved, reaction rates are not affected. Therefore, the aim of mixing is not to produce a homogeneous suspension per se, but to ensure that an unequal solids distribution in the mixed or aerated zone does not significantly limit conversion rates.

This chapter will present some methods to quantify homogeneity or mixing. However, the difference between well mixed and quality mixing (from a microbial point of view) is not yet understood. Floc break-up, hindered slip velocity zones, measurement limitations, etc. are impeding knowledge of connections between maximum mixing (either measured by the coefficient of variation (CoV), solids variation or CFD) and removal/conversion rates.

3.2 QUANTIFYING THE DEGREE OF MIXING

To provide good mixing conditions or adequate mixer design, one issue must be addressed: how to measure the level of mixing. However, the question of how to quantify a 'homogeneous suspension' in aeration tanks is the subject of an on-going debate as other authors have defined different benchmarks for this condition. Different methods have been proposed for quantifying how well a suspension is mixed. These can be divided in two main groups: traditional methods (see Sec. 3.2.1) and those embraced in a new generation of methods: Computational Fluid Dynamics (CFD) which are trying to combine hydraulic process models with process biology (see Sec. 3.2.2).

In the majority of designs the mixing conditions are predicted via the ideal reactor model (Stamou, 2008) and a completely mixed flow regime is assumed. The reality, however, is that many factors can impact this theoretical 'good or complete mixing' in the AS process. Even if some common rules of thumb indicate the minimum air requirements to provide good mixing conditions (Metcalf & Eddy, 2003; Samstag *et al.*, 2012), none of them can assume the impact of other factors. Neither provide information if these minimum requirements can be improved or optimized depending on each scenario or given conditions.

Is commonly accepted that the air requirement to ensure good mixing will vary from 1.2 to 1.8 $m^3\,h^{-1}\,m^{-3}$ of tank volume; and typical power requirements for maintaining a completely mixed flow regime with mechanical aerators varies from 13 to 26 W per m^3 of tank volume (Karpinska & Bridgeman, 2016). But as we mentioned none of these assumptions consider the impact of tank hydraulics (cross-section, depth or presence of baffles), energy input, or any other variable sensitive to affect mixing. In this sense, some authors stated that when tank geometry, tank volume, agitator/mixer position and aeration are favourable, volume-specific mixing energy can be reduced (Füreder *et al.*, 2018). In the following sections more details will be provided.

3.2.1 Quantifying mixing: traditional methods

A few of the most common methods used to quantify mixing are listed below. Some of them are more popular or used than others, and all of them have their own limitations:

- **Coefficient of Variation (CoV)** or the ratio of the standard deviation in solids concentrations to the average solids concentration.
- The **maximum variation of solids** concentrations (difference between minimum and maximum concentrations) often as expressed as a fraction of the average solids concentration.

- The **height of the cloud** of the suspension in a vessel, often relative to the depth of the vessel, or inversely, the height of clear liquid above the suspension.
- Determining the point where additional mixer power does not improve reaction or **conversion rates**.

3.2.1.1 Coefficient of Variation (CoV)

From the previous list of the most common methods intended to provide some sort of measurement of the level of mixing, the coefficient of variation (CoV) is considered by some engineers the most repeatable and relevant. The CoV, also known as relative standard deviation, is a standardized measure of dispersion of a probability distribution. It is often expressed as a percentage, and Cekinski *et al.* (2010) defined CoV as the ratio of the standard deviation to the mean of the suspended solids concentration in a reactor:

$$\text{Coefficient of Variation (CoV)} = \frac{\text{Standard Deviation}}{\text{Mean}} \tag{1}$$

The lower the value of CoV, the better the mixture quality. Although CoV is one of the most understood and used methods for measuring the level of mixing, there are still important limitations in the current understanding when applied to biological processes. Unfortunately, CoV only provides a general measure on the distribution of suspended solids in the tank, but do not provide direct measure or any information about the quality of the flocculation or the impact on microbial activity (or conversion rates).

Good mixture (i.e., MLSS) distribution or homogeneity may result in low CoV, but good mixing indicators can come at the expense of excessive turbulence or shearing for the sensitive microbial population. Excessive mixing, especially by the use of impellers, may be detrimental for the formation and/or structural integrity of microbial flocs, as they will be potentially sheared and disaggregated, producing a poorly settling sludge.

In the chemical industry it is usually suggested that a CoV less than 25% can be considered complete suspension, and that further increases in impeller speed achieve little improvement in homogeneity. This criterion is less stringent compared to 10% often used in mixer specifications in the wastewater industry.

Nevertheless, the maximum CoV that would result in a significant reduction in the conversion rates (or nutrient removal rate) is not known at present. Combined CFD and kinetic models are being developed in order to address the issue. At present, a CoV of approximately 5% seems like a reasonable target that is readily achievable, without causing an obvious negative effect on reaction and conversion rates. There is some evidence that maintaining higher CoVs are compatible with acceptable performance, such as the 33% target used at Milwaukee Metropolitan Sewerage District (S. Arora, 2017, personal communication).

Uncertainty on how CoV results are obtained also hinders its reliability. The CoV results would depend on the number of samples collected, as well as how close to the surface is the first sample collected and how close to the floor is the last one. Therefore, for CoV to be universally comparable, a standard needs to be developed, specifically regarding the level of the first sample (how close to the floor?) and the last (how close to the surface?). CoV appears to be particularly sensitive to clear zones at the surface and to make full use of the available zone volume such clear zones need to be avoided.

3.2.1.2 Maximum variations of solids

The maximum variation of solids (the difference between the maximum and minimum concentrations) is also one of the most commonly used criteria for 'good mixing' in AS process (Samstag & Wicklein, 2014; Schauer & Kobylinski, 2010). The variation of solids approach consists in collecting samples from various locations and depths to obtain an understanding of solids concentration distribution.

The accepted criterion for 'good mixing' in AS process control is that variations of solids concentration across the complete mixed tank profile should be less than 10% (Samstag & Wicklein, 2014). However, the proper design of such 'well mixed' AS systems should require a further sophisticated analysis of reaction rates (such as OUR), flow hydrodynamics for mixing patterns in the tank, distribution of the oxygen, etc.

Currently both maximum variation and CoV are used to quantify mixer performance. While maximum variation would tend to be much higher for a given data set than CoV, it suffers from the fact that a single sample, in particular an outlier, can determine the final result. For both CoV and maximum variation standards are required for the number of readings that needs to be collected, as well as the distance from the surface and the floor the first and last readings need to be taken, respectively.

3.2.1.3 Height of the cloud

A clear liquid layer at the surface of a reactor zone is seldom observed in activated sludge systems, unless a mixer system has failed, or has been turned off. For this reason, methods involving measurements of the cloud height or the depth of the clear liquid is seldom used in the wastewater industry. The increased use of ballasts, currently either sand or magnetite, to significantly increase sludge settleability, may eventually make these methods relevant to the industry. If it does, a significant body of research, developed by the chemical processing and mining industries, stands ready to be used.

3.2.1.4 Reaction rates improvement/Biological activity indicators

The characterization of aeration basins, ponds or similar systems by means of vertical and horizontal profiles that directly monitor biological conversion rates may benefit our understanding of the existing mixing conditions. Traditionally, the limits of our understanding mixing conditions simply rely on measuring the variations in concentration of some parameters (mainly TSS) along the tank.

The aforementioned CoV method can be used not only to total suspended solids (TSS) but also to other key parameters, ranging from contaminants (i.e., BOD_5, nitrite, and nitrate) to process indicators, which can bring satisfactory insights on the existing homogeneity conditions. While vertical profiles regarding pollutant concentrations can deliver little information, other two key parameters are more suited in terms of providing a better picture of the distribution of those reactions unfolding within the reactor tank. In this section the use of parameters as DO, ORP, temperature, pH and OUR as mixing indicators relating to conversion reactions will be discussed. The use of some of these is discussed below.

DO and ORP mapping

For example, Figure 3.1 presents the result of a study (Jiang *et al.*, 2018) that combined both DO and ORP measurements to determine if different sections/areas of the tank were presenting the expected conditions (i.e., anoxic, aerobic, anaerobic) and how close they were from the ideal (or theoretical optimum) performance.

These types of results combined with process characteristics (and expected conditions to comply with effluent targets) can deliver valuable information about the process, or can also confirm the consistency in any previous measurements using TSS CoV. Variations in ORP and DO values can provide a comprehensive level of detail on the reactions unfolding at a given time and place. Nevertheless, this approach can often present some bias or inaccuracies on the measurements as result of equipment limitations. Jiang *et al.* (2018) showed that ORP may appear lower than its acknowledged definition for each of the microbial metabolisms due to the response time of the probe electrode. Luminescent DO (so called LDO) probes tend provide a low-pace response in comparison to the electrochemical (or membrane) probes, which may hinder the results' interpretation, and different response considerations need to be taken into account.

Moreover, using DO measurement as indicator of mixing may have also some limitations, especially about low limit DO values. Low DO values (i.e., DO $<$ 0.5 mg/L) is hard, if not impossible, to define the actual oxygen conditions (aerobic, anoxic, anaerobic) and certainly to discriminate between the anoxic and anaerobic metabolisms, due mainly to the noise-to-signal ratio of DO probes at their low range of operation. Other factors, such as biological conversion rates, substrate concentrations, concentration of various electron acceptors, etc. also affects whether anoxic or anaerobic reactions take place.

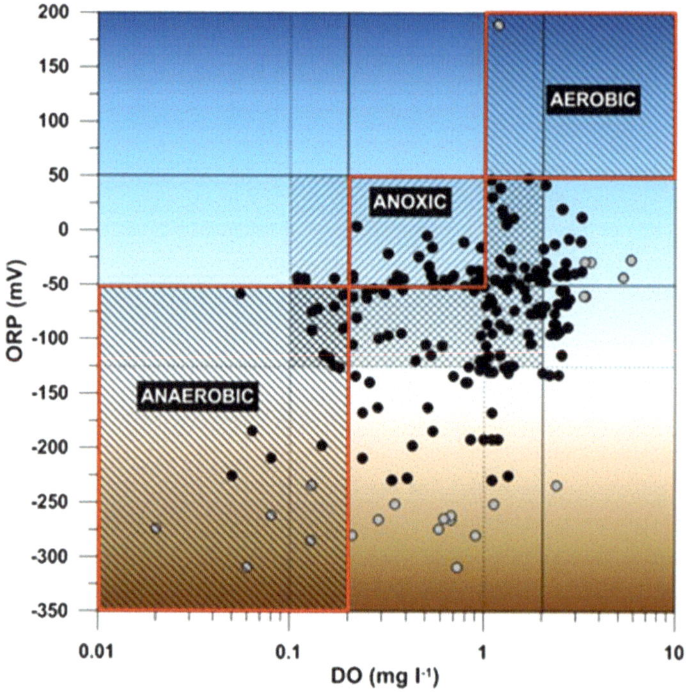

Figure 3.1 Dissolved oxygen plotted against the oxidation-reduction potential (ORP). The red rectangles represents potential regions where anaerobic, anoxic, and aerobic conditions are defined according to DO. The shaded areas represent the definitions based on ORP. Outliers are plotted in light grey (Jiang *et al.*, 2018).

OUR mapping

One of the promising approaches in the monitoring of mixing is tank mapping by using oxygen uptake rate (OUR) measurements. OUR is an indicator of biomass activity (mg O_2/l/h) and can became an important technique in assessing the distribution of the biological reactions taking place in the aeration tank. The aforementioned disadvantages in the interpretation of the TSS, COD or other parameters measurements would encourage the use of respirometric methods for assessing the homogeneity of the kinetic constants along the tank.

Recently, it has been proven that these limitations in mixing can be easily detected with onsite OUR measurements. Two main methods can be used depending on the expertise, budget and complexity: direct or indirect. The most exact method is the use of direct in-situ respirometric OUR (EPA, 1989), however the most practical method yielding also reliable results is the use of indirect OUR calculations by using off-gas measurements (see Sec. 2.3.2). Combining the oxygen transfer efficiency (OTE, %) measurements and the air flux supplied in the system ($m^3\ m^{-2}\ h^{-1}$) how much of this oxygen has actually been used by the microbial population can be calculated. This measurement is a direct indicator of the microbial activity (pollutants mineralization or conversion rates). The calculation, and corresponding mapping, of the microbial activity distribution by OUR can detect those areas where the suspended biomass is receiving less load due to hydrodynamic/mixing issues. These OUR readings will help to identify those areas where the values are lower than expected in comparison neighbouring areas or past experience. However, we should clarify the main limitation of this approach, which consists in its bidimensionality. The bubbles collected represent the average of the gas transfer phenomena over the water depth, and there is no depth mapping when bubbles are collected at the surface. In theory the off-gas could be collected below the surface, thereby allowing the mapping in 3-D, but the practical limitations of pushing the gas hood in depth, combined with the lower gas transfer at lower depths (while the instrument error remains unaffected by depth), severely limit this approach. OUR mapping via off-gas relies on the collection of off-gas at the surface, which means that the bubbles reaching the surface at the point of collection reach the analyser. These bubbles may have been generated below the off-gas hood or in its vicinity (especially upstream, due to the average forward liquid velocity). Hence, sampling uncertainty is inherently built into this measurement.

Figures 3.2 and 3.3 were created to show two typical OUR profiles in aeration tanks: while Figure 3.2 shows a quite homogeneous distribution of the microbial uptake of oxygen across the width of the basin, in Figure 3.3, in the other hand, it can be seen how the OUR mapping unveils potential shortcuts in the

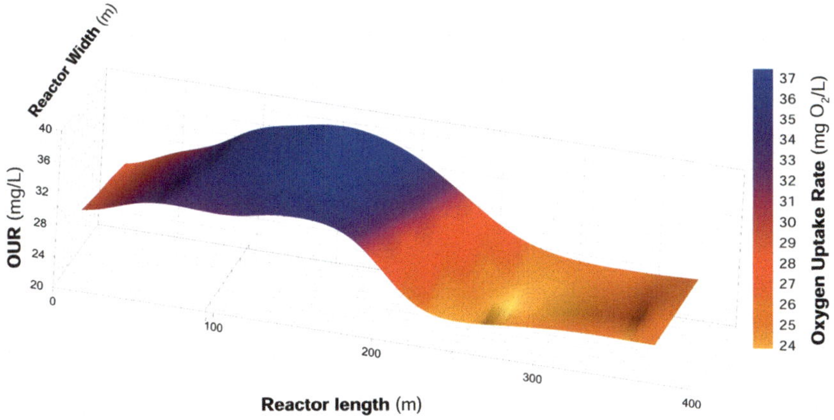

Figure 3.2 OUR mapping. The X axis corresponds to the aeration tank length, Y axis to the oxygen uptake rate ($mg_{O2}/l/h$) and Z to the width of the reactor pass. Note the limited gradient across the reactor width.

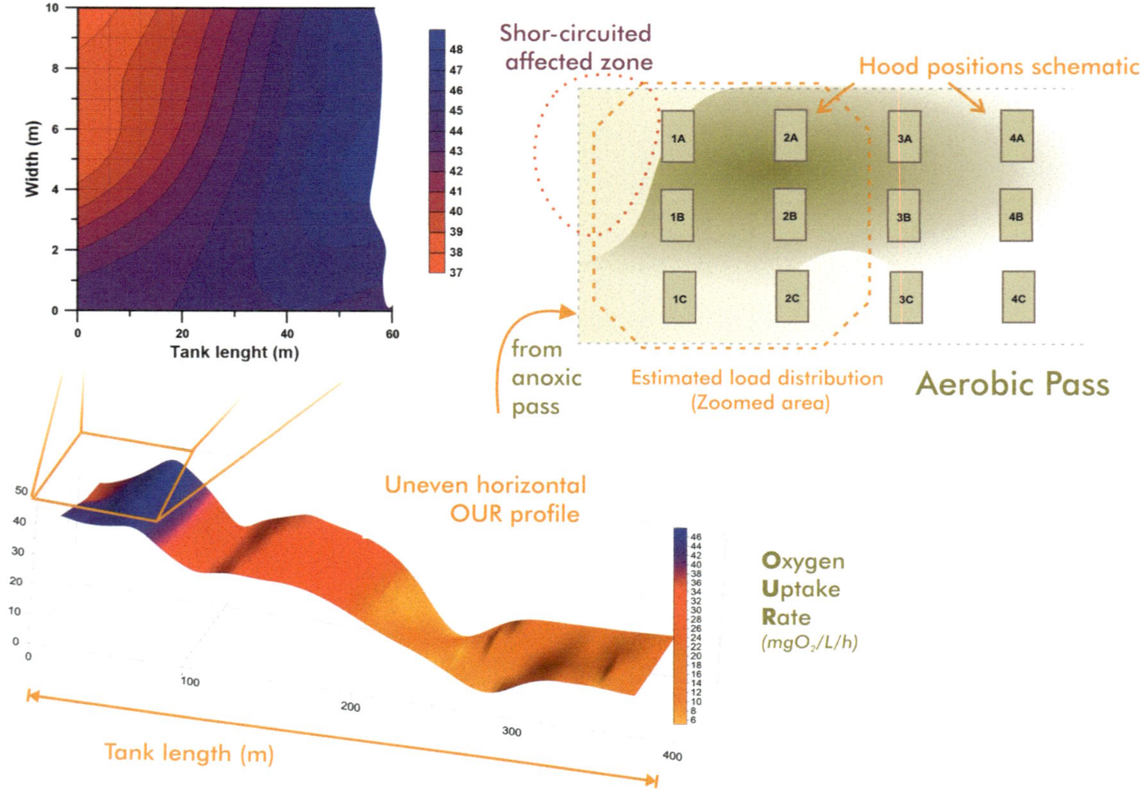

Figure 3.3 OUR mapping. The upper figure shows the corresponding position of the deployed hood to retrieve the off-gas measurements (OTE, %). The X axis corresponds to the aeration tank length, the Y axis to oxygen uptake rate (mg_{O2}/l/h) and Z to the width of the reactor pass. Note the visible gradient across the reactor width.

hydrodynamics which are decreasing the reactive or functional space within the tank. The maximum conversion rate potential could be recovered by hydrodynamics deflectors, such as a baffle with equally-spaced openings, or by improving lateral mixing at those specific areas.

As indicated, this method can identify mixing issues not only related to insufficient blending of different feed flows or unequal distribution of biomass and/or substrate, but by preferential flow patterns. This will potentially highlight the accuracy of an implicit assumption prevalent in the industry today: that there are no preferential flow paths, or short-circuiting, in existing basins.

Temperature

While a detailed temperature analysis could eventually pin point some mixing limitations is important to note that temperature alone is not a sufficient condition to define thorough mixing, due to the specific heat capacity of water. In combined sewer systems the inflow temperature changes during rain events can be interpreted as alternative tracer experiments and has the potential to identifying hydraulic characteristics (Ahnert *et al.*, 2010). Besides, it may prevent misuse of calibration in over-parameterised models, for example, activated sludge models. The results from the previous author identified that dead zones were about 10–15% of the reactor volume. Although these amounts are small, they are statistically significant. The same can be concluded for pilot-scale tests were the active volume fraction is much smaller with values around 0.7 and a short-circuiting flow of about 50% of the inflow.

Nevertheless, temperature is not the only conceivable alternative tracer measurement. In some special cases, ammonia or nitrogen can also be used (e.g., Braun & Gujer, 2008). Under the assumption of no conversion processes in primary or secondary settlers, the measurements of sufficiently fast online probes in influent and effluent can be used for a similar evaluation. If the conversion processes are integrated into this methodology, reactive tracers may also be used.

3.2.2 New generation: methods for mixing quantification

Existing process models fail to determine the minimum degree of homogeneity above which the distribution of suspended solids does not significantly affect the overall process performance. New methods are required to relate the variation in solids concentration with the existing biological conversion rates. Therefore, current process models simply assume completely mixed conditions (i.e., perfect homogeneity) in a reactor element and cannot address this question. Although the issue has not been successfully addressed, it is the subject of on-going research.

In the last few years, the use of Computational Fluid Dynamics (CFD) has evolved into a promising technique for design, optimization and control of the

AS systems. A next generation of process models, which combine CFD analysis and biological reaction rates is currently being developed (Laurent *et al.*, 2014; Rehman *et al.*, 2014) and will attempt to overcome current limitations. Potential possibilities of coupling of the CFD data with ASM based codes have been explored in order to generate a suitable tank-in-series model, where the biokinetic model can be implemented (Karpinska & Bridgeman, 2016). However, to date, a model capable of a complete three-phase CFD model coupled with ASM1 aiming to quantify the transport and mutual interactions between water-gas-sludge phases within the AS reactor has proven elusive. Despite providing more reliable results than those obtained from ASM, this approach relies on the trade-off between a model which can be implemented and run in a realistic timeframe, and the extent of simplifications and the solution accuracy, which may lead to a series of output errors (Karpinska & Bridgeman, 2016).

CFD modelling has evolved over the last decade as a valuable tool to assess solid-liquid interactions in different reactor design configurations. In order to obtain accurate results simulating activated sludge mixing as a function of mixing system and energy input, CFD models must incorporate a source term for solids settling and transport that couples the influence of solids gradients with the density profile and fluid flow in the reactor as described by Samstag *et al.* (2012). In the CFD modelling, hindered settling velocities are usually calculated based on value for the sludge volume index (SVI), using the revised Daigger equation (Daigger, 1995), allowing a mixed liquor suspended solids concentration to be determined by the coefficient of variation.

However, the complete CFD simulation of the complex multiphase flow in AS tanks remains a challenge, due to the intense computing requirements and limited feasibility resulting from the imposed convergence criteria. Although there is still no unequivocal protocol on CFD methodology, some models (i.e., RANS/URANS closed by sk−ε turbulence model) are being adopted as the standard for the modelling. Further details on CFD and their current limitations and on-going research can be found in the review by Karpinska and Bridgeman (2016).

3.3 DESIGN CRITERIA

3.3.1 Comparison criteria for mixing methods

A number of different approaches have been proposed to compare mixing alternatives and help during the scale-up process from laboratory (or pilot scale) to full scale, or for designing mixers.

The rules of thumb based on past experiences and years of practice are the most used guidelines when trying to establish comparisons between mixing methods and to determine the mixing requirements in WRRFs. Mixing (or mixed liquor flow velocity induction) can be obtained through different methods (see Sec. 3.4). Using or manipulating airflow rates, effluent flows, pumps or impellers can deliver the required mixing to satisfy flocculation and the corresponding conversion rates. The balance between proper mixing in order to maintain the required level homogeneity need to be in accordance to bioflocculation, as stronger or unbalanced mixing could deliver undesired break of floc or inadequate conditions for microbial growth.

3.3.2 Mixing design criteria: rules of thumb

In the wastewater industry, various rules of thumb have traditionally been used. The reader is strongly advised against the indiscriminate use of such rules, in particular those that do not translate into a constant power level (including airflow per unit volume and airflow per unit length or channel). Nevertheless, popular due to their simplicity, rules of thumb are widely used, helped by the perception that these rules treat different mixer configurations and suppliers the same. Table 3.1 summarizes typical design guidelines used in the industry.

Design criteria listed in Table 3.1 are conservative guidelines originating from a generalization of specific full-scale design and performance experiences. While adopting such published design guidelines for other mixing designs is typically safe for the design engineer, the drawback is that these guidelines are based on worst-case scenarios and – as a result – tend to increase over time.

Füreder *et al.* (2018) stated that when the tank geometry, tank volume, agitator position and aeration are favourable, mixing energy can be about 24 W/m^3. Similarly, other authors maintain the typical power requirements for a completely

Table 3.1 Recommended mixing criteria commonly cited in literature.

Application	Units	Values
Fine-bubble aeration	$scfm/ft^2$ $Nm^3/h \cdot m^2$	**0.12** **2.2** *(WEF MOP 8, 2010)*
Coarse bubble in spiral roll configuration	$scfm/1000\ ft^3$ $Nm^3/h \cdot m^3$	**20–30** **1.2–1.8** *(WEF MOP 8, 2010)*
Channel aeration	scfm/linear ft $Nm^3/h \cdot m$	**2–5** ***11–28*** *(WEF MOP 8, 2010)*
Mixer power level in unaerated zones	hp/Mgal W/m^3	**30–40** **6–8, 14** *(WEF MOP 8, 2010)*
Velocity gradient (*G*), optimum flocculation – minimum tested	s^{-1}	**20–75** (*WEF MOP 8, 2010*) **40** (*Parker et al., 1971*) **15** (*Wahlberg et al., 1994*) **10** *(Parker et al., 1971)*

Table 3.2 Comparison of different mixing criteria.

Source	Power Level Expressed as		
	$Nm^3/h \cdot m^2$ ($scfm/ft^2$)	W/m^3 ($hp/10^6$ gal)	s^{-1}
Fine bubble aeration	2.2 (0.12)	(30)	77
Coarse bubble (20 scfm/kcf)			
SWD = 13 ft (4.0 m)	4.8 (0.26)	13 (66)	114
SWD = 20 ft (6.1 m)	7.3 (0.40)	20 (100)	141
Channel aeration (2 scfm/ft)			
W = 5 ft (1.5 m)	7.3 (0.40)	20 (100)	141
W = 15 ft (4.6 m)	2.4 (0.13)	6.6 (33)	81
Mixer power in unaerated zone	2.9 (0.16)	8 (40)	89
	5.0 (0.28)	14 (70)	117
Velocity gradient, *G*	0.15 (0.008)	0.4 (2)	20
Optimum flocculation	0.59 (0.032)	1.6 (8)	40
	0.083 (0.0045)	0.23 (1.2)	15
Minimum tested	0.037 (0.002)	0.1 (0.5)	10

mixed flow from 13 to 26 W per m^3 of tank volume (Karpinska & Bridgeman, 2016). Other broadly referenced power levels as the ones found in Metcalf and Eddy increased from 13–26 in 1972, to 15–30 in 1979, to 19–39 in 1991 to 20–40 W/m^3 in Metcalf and Eddy (2003, 2014). This is progress of the wrong kind, suggesting that the progress our industry has made during the past three decades are mixing systems that are ever less efficient. In addition the final range is well above the range listed in Table 3.2. On the other hand, the 6th edition of WEF MOP 8 (2017) is less prescriptive and includes mention of the concepts discussed here.

Another drawback of this prescriptive approach is that it creates an incentive for the equipment supplier to furnish an inefficient unit that uses a lot of power. This approach actually precludes an innovative, energy-efficient design from meeting the project specifications. It is also based on the assumption that the specifying engineer has a better understanding of mixing technology than the equipment supplier, a condition that rarely, if ever, occurs.

At first glance, it is confusing that mixing the same material (mixed liquor) requires such different energy requirements and that different technologies are designed on basis of various power units that cannot intuitively be compared with each other. However, notice that the units for fine bubble aeration, mixer power in unaerated zones and the flocculation requirement are all expressed in equivalent units of power per unit volume, or power level.

Table 3.2 compares the different design criteria listed in Table 3.1 in terms of the same units, power level. Typical system dimensions have been assumed to compare the power levels for different mixing technologies.

The table shows that mixing requirements *not* based on power level are strongly affected by tank dimensions: using the coarse-bubble guide, expressed as an airflow per unit volume, would increase the power level in direct proportion to the side water depth (SWD) of the tank. Likewise, using the channel aeration guide, expressed in airflow per unit length, would result in high power levels for narrow channels, but that power level decreases in inverse proportion to channel width. This suggests that these criteria exist only for legacy reasons, have no basis in the underlying principles or fundamental analysis and are best avoided. Notice that for both criteria, many of the mixer power levels cited in Table 3.2 exceed the widely used values in the first line by a factor of three or four.

There is good agreement between the requirement for fine bubble aeration and mixer power in unaerated zones. This implies one of two possibilities: (1) both values are accurate and close to optimum, or (2) one value was originally calculated using the other.

The issue is highlighted by the much lower power levels that were experimentally determined by Parker *et al.* (1971), as well as Wahlberg *et al.* (1994). In their cited studies, the objective was achieving optimum flocculation, not minimum mixing requirements. However, presumably the mixed liquor solids need to stay in suspension for flocculation to occur. In other words, the optimum power level for flocculation cannot be lower than the minimum power level required for mixing, but the minimum mixing requirement may be much lower than the optimum power level for flocculation. The flocculation results suggest that optimization would allow mixing power to be much lower than indicated by traditional design guidelines. It should be noted that the flocculation results were determined under ideal laboratory conditions, and may need to be adjusted at full scale, to allow for decay of power level over distance, obstructions and other factors.

3.3.3 Mixing efficiency and power requirements

This section presents the compiled results of the different cases studies embracing the four mixer categories (i.e., mixing from impellers, pumps, air, and hydraulics). The results were obtained by three independent groups of researchers, showing remarkably similar results. The following figure shows the relation between power requirements and coefficient of variation (CoV) for the four different typologies of mixing discussed in previous sections (i.e., mixing from impellers, pumps, air, and hydraulics). The results were obtained by three independent groups of researchers, showing remarkably similar and consistent results.

From Figure 3.4 the first thing to notice is that the only mixing method that stands out is the pump or jet mixer, which required power levels of around 20 W/m^3 (100 hp/mil gal) to achieve a CoV of 10% or lower (see case study

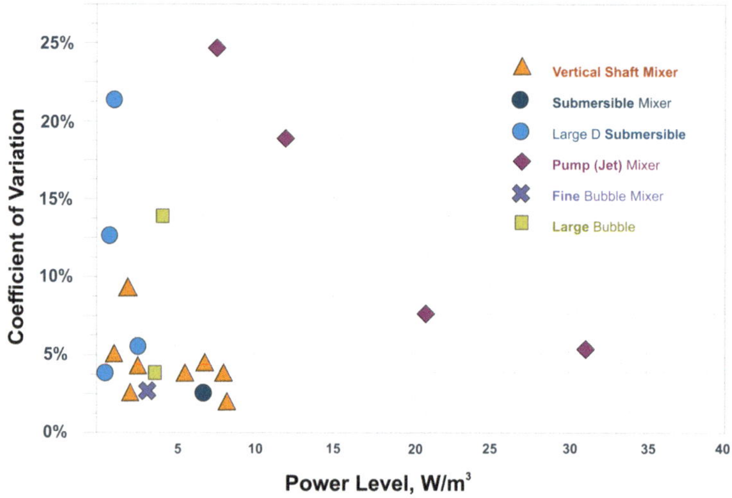

Figure 3.4 Impact of power level on coefficient of variation (CoV).

below). This may be related to scale (diameter and depth of the tank involved) as well as the fact that the directional mixer produces relatively little fluid velocity in the vertical direction so that little fluid velocity is achieved at the surface of the tank. Some power is also consumed by head loss through the piping system and this head loss does not contribute to mixing.

The next three outlying CoV values (submersible mixer at 6.5 W/m^3, large bubble mixer at 3.6 W/m^3 and, less obviously, the vertical shaft mixer at 1.1 W/m^3) were all produced in tanks where different streams were blended. This strongly suggests that combining different feeds before introducing them into a mixed tank can reduce CoV (which means improved mixing) and allow lower power levels to be used. Upstream blending would also typically eliminate the need for mixing in a horizontal direction, making plug-flow designs more viable.

It can be seen that the rest of mixing equipment reported values below 6% CoV until the power level drops below 1 W/m^3, and for some applications much lower. These results show limited dependence of CoV on power level, with CoV decreasing as power level is increased. Nonetheless, the dependence is not very strong, suggesting that increasing the power level for the submersible mixer from 2.6 W/m^3 (13 hp/mil gal) to 7.7 W/m^3 (39 hp/mil gal), that is, by a factor of three, would reduce the CoV from 3.9% to 3.0%, or a relative reduction of 23%.

The question of how low a power level is sufficient to keep mixed liquor in suspension has not been definitively answered yet, and may never be answered in a general sense. A specific answer for each application may need to be developed. However, the data suggest that vertical shaft mixers, or at least one kind of impeller, can achieve low CoV of 5.5% at a power level of 0.1 W/m^3 (0.5 hp/mil gal). It is also possible that further reductions in power levels can be achieved while maintaining acceptable CoV values. Thus, the results presented suggest a power level of 0.1–0.2 W/m^3 (0.5–1.0 hp/mil gal) may be sufficient to keep mixed liquor in suspension. Although the general applicability of these low power levels remains to be confirmed, the potential for further reductions in power levels has not been excluded. To deal with these uncertainties, the appropriate approach is to design mixers with adjustable power levels, so that operators can optimize systems once they are in operation.

To illustrate the difference between the traditional rules of thumb and a design optimized to minimize power use and optimize flocculation, consider the following example: a set of five activated sludge reactors consist of three identical anoxic zones ($L = 10$ m) followed by two aerobic zones ($L = 20$ m). Each reactor has a width of 15 m and a side water depth of 6 m. The mixed liquor is transferred to the secondary clarifiers by a channel with total $L = 200$ m, $W = SWD = 2$ m. Waste activated sludge is stored in a sludge holding tank with $L = 20$ m, $W = 15$ m and $SWD = 6$ m. Determine the mixing requirements or minimum aeration rates.

The rule-of-thumb minimum air flux for aeration in the aerobic zones is 2.2 Nm^3/h · m^2 (Table 3.2), which corresponds to an airflow of 660 Nm^3/h (=2.2 × 20 × 15). The airflow in the sludge holding tank would be 2160 Nm^3/h (=1.2 Nm^3/h · m^3 [at the low end of the range for coarse bubble aeration mixing in Table 3.2] × 20 × 15 × 6). The airflow required for channel aeration would be 2200 Nm^3/h (=200 × 11 [low end of the range for channel aeration in Table 3.2]). Table 3.2 offers two values as optima for flocculation, $G = 15$ and $G = 40\ s^{-1}$. Without getting into which value may be more generally applicable, the higher value will be used in this example, which corresponds to an air flux of 0.59 Nm^3/h · m^2. For the aerobic zone this corresponds to an airflow of 177 Nm^3/h (=0.59 × 20 × 15). For mechanical mixers in the anoxic zones, a minimum installed power for the rule-of-thumb is 5.4 kW (=6 [low end of the range in Table 3.2] × 10 × 15 × 6/1000; ignoring limitations imposed by available motor sizes), while it is 1.4 kW (=1.6 [Table 3.2] × 10 × 15 × 6/1000). The results for all the mixing systems are summarized in Table 3.3.

Some of the most significant differences occur for conditions where the traditional criteria are not based on consistent power per unit volume values (holding tank and channel aeration). The difference would be even more significant for a deeper holding tank or a channel with a smaller width. The channel presents a unique opportunity to flocculate the mixed liquor upstream of the secondary clarifiers.

In Table 3.3, the total airflow is calculated based on the following assumptions:

- The minimum airflow per zone is assumed to apply continuously to the second of the two aerobic zones in each reactor. Whether or not this is

Table 3.3 Comparison of traditional and optimal methods for sizing mixers.

Mixing Criteria for Example	Traditional	Optimal	Reduction
Minimum Airflow per Zone, Nm^3/h	660	177	73%
Holding Tank Airflow, Nm^3/h	2160	177	92%
Mixed Liquor Channel Airflow, Nm^3/h	2200	236	89%
Anoxic Zone Mechanical Mixer, kW	5.4	1.4	73%
Total Mixing Airflow, Nm^3/h	6580	1210	82%
Total Mixing Power, kW	222	48	79%

the case will depend on many factors including system SRT, loading rates, etc. Also note that due to the large difference between the two minima, it is possible that the basin using the traditional minimum criterion would be the only one operating at the minimum airflow. The design engineer must determine whether the minimum airflow applies or not, using process models and aeration airflow calculations described elsewhere in this book.

- The holding tank airflow shown is just the estimate needed to maintain the solids in suspension, in part explaining the large potential savings. Depending on the operating SRT, the WAS may need more airflow to avoid exhausting the available oxygen and creating septic conditions. Again, due to the large difference between the traditional and optimal requirements, the airflow required to maintain DO may be in between these two values. One strategy for operating holding tanks or aerobic digesters is to cycle the air on and off to switch between aerobic and anoxic conditions. To calculate the total mixing airflow, it was assumed that the airflows shown would be used 50% of the time.
- Major savings are also available for channel aeration. This is typically an overlooked opportunity to flocculate the mixed liquor, especially for channels transferring mixed liquor from the reactor to the clarifier. It is also worth considering whether the flow velocities in these channels are sufficient to keep the solids in suspension and achieve flocculation. *At a plant in the United States, a primary effluent channel was taken out of service to install a mixer system after 40 years in service. Observing that about 10 cm of solids had accumulated in that time, the operations superintendent asked: 'How much energy are we going to spend to avoid an inch of solids settling every decade?' And that's assuming the new system will completely prevent solids from settling …*

The total mixing power estimate depends somewhat on blower design assumptions, including factors such as site elevation, dynamic head loss the air piping, etc. as discussed elsewhere in the text.

The choice of velocity gradient, $G = 40\ s^{-1}$ is arbitrary in this example, and the user may choose a different G value. To use a different G value, it is important to remember that when calculating G, the square root of the power level and air flux values are used, as a quick review of the numbers in Table 3.3 will confirm. Choosing a value of say $G = 25\ s^{-1}$, would change the design air flux to 0.23 $Nm^3/h \cdot m^2$ ($=0.59 \times [25/40]^2$).

This raises the important issue of identifying the optimum velocity gradient that should be used for a given design. For existing systems, the parameters describing flocculation can be determined experimentally using the method described by Wahlberg *et al.* (1994). These parameters can then be used to optimize the design for flocculation. For new systems, the opportunity to determine these parameters adds to the case for operating a pilot plant as part of the design project, as urged in other chapters (see Chapter 8). Given the limited research and data currently available, the design engineer is encouraged to incorporate flexibility into the design so that operations staff can make adjustments in the field to correct for unknowns, including limits in our current understanding of flocculation, or changing conditions, which may affect flocculation parameters.

At many facilities the presence of small solid particles in the secondary effluent appears to be the result of high turbulence and shear in the conduits used to transfer the mixed liquor from the activated sludge reactor to the secondary clarifier. In some cases, this is hard to avoid, for example if the site has a strong slope it would be difficult for the designer to avoid high head loss between the reactor and the clarifier. In too many cases, though, the high turbulence appears to result from choices made by the designer when sizing the conduits. All the efforts invested in designing a reactor for optimum flocculation will not materialize if the floc is subsequently sheared during transfer to the clarifier. To determine if the flow conditions anywhere during transfer from reactor to clarifier may result in floc shearing, the methods described by Das *et al.* (1993) can be used to correlate power loss during transfer to a velocity gradient value.

3.4 MIXING EQUIPMENT

Different mixing equipment and strategies are available to induce liquid flow velocity, which is needed to maintain the mixed liquor in suspension. The integrity of sensitive flocs or microbial aggregates needs to be preserved to avoid negatively impacting sludge settling characteristics. The most typical methods for providing the required velocity include:

- using rotating impellers to induce such velocity,
- using pumps to withdraw mixed liquor and return at velocity,
- releasing gas, usually air, close to floor level, and
- using the flow of effluent under the influence of gravity

In the rest of the chapter the different mixing technologies commonly used in the wastewater treatment industry will be reviewed. First, rotating mixing equipment will be considered first, and then pumped mixing, also known as jet mixing. The scenarios in which mixing and aeration as well can be obtained by air-powered equipment are described. These mixers may also be used to supply oxygen to the process, in which case the design engineer must consider both mixing and oxygen transfer requirements. Finally, flow-induced mixing will be considered.

3.4.1 Rotating impellers

Rotating impellers can induce flow in either a radial or an axial direction (Figure 3.5). In wastewater applications, axial flow impellers are most commonly used. The axial flow can be forward (away from the shaft through which the impeller is powered) or backward. The chief example of backward flow would be up-pumping vertical impellers. The main exception to axial flow would be the hyperboloid mixer (Figure 3.6) that is designed to induce flow radially from a vertical impeller, specifically along the floor of the zone.

The orientation of the impeller axis is important. Both vertical and horizontal impellers induce significant horizontal flow velocity, with the vertical impeller arguably producing a more uniform distribution of such velocity. A vertical impeller requires structural support as well as access to the motor, which is typically located on a platform above the basin. This means that choosing a vertical impeller has implications for the structural design of the basin, and that switching to a vertical impeller at a later date may not be practical. Floating vertical impellers may also be used to avoid structural requirements.

Horizontal impellers, on the other hand, usually require a submerged mixer motor, since wall-penetrating impellers are seldom used in the wastewater industry. Submerged motors have proven to be challenging to keep free of leaks. In fairness to suppliers, some issues have been related to maintenance, more than design or manufacturing.

In terms of performance, horizontal impellers are usually attached to zone walls, which means that flow to the impeller may be restricted by the wall. The directional thrust of such flow, could also make it difficult to distribute the velocity coming from such an impeller evenly throughout the area of the zone, so that the direction must be carefully considered. An adjustable arrangement could be used to allow the operators the flexibility of adjusting the direction.

Impeller design is an area of on-going research and development. Original impeller designs were transferred from the chemical engineering industry and were prone to ragging. While the industry appears to be moving towards improved and finer screens, assuming that water is free of any rags would not be wise for the foreseeable future. Fortunately, non-ragging designs are available and should be used.

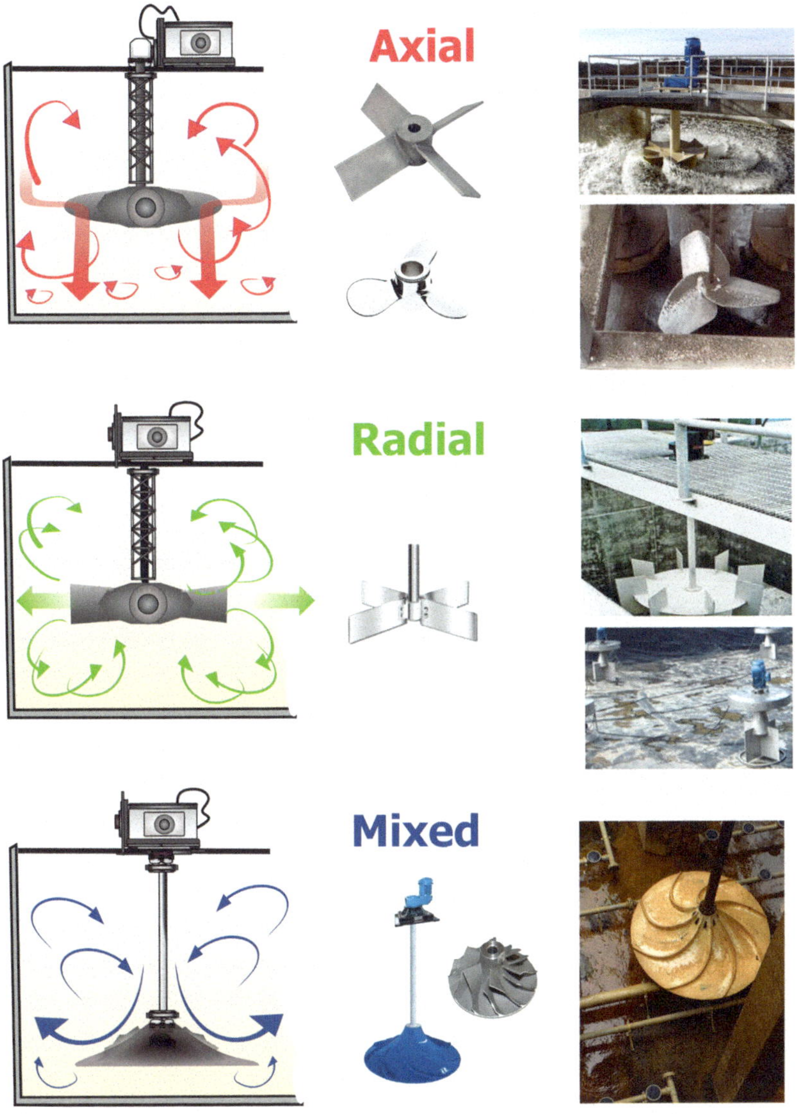

Figure 3.5 Flow direction of three different pumps/impellers.

3.4.2 Pump mixers

Sometimes, mixed liquor can be cycled through an external pump and returned the mixed zone. In these systems, there are head losses that add to power use without contributing to mixing, such as head losses associated with flow in conduits, as well as losses from elbows, tees, valves, and flow entry. The combination of

Figure 3.6 Pictures of the hyperbolic mixer deployed at a WRRF, installed above the existing diffuser grid.

these head losses cause the relatively high power demands (that is, relatively low efficiency).

Pump mixers, draw the mixed liquor through a pump before discharge into the mixed zone, which can take the form of any number of arrangements, but typically via a nozzle where the aim is to create a high velocity jet, from which jet mixers take their name. The jet induces mixing as it dissipates in the mixed zone. The high shear to which the mixed liquor is subjected as it passes through the pump can contribute to deflocculation. Jet mixers are often designed with an air supply, for use as a jet aerator. This is a popular arrangement in sequencing batch reactors where the jet mix system can be used during both unaerated (no air supply) and aerated phases. The bubbles transit through the nozzle and are released in the jet zone at a velocity higher than their terminal velocity, and dissipate their energy by decelerating and opening in a plume where bubbles rise at their terminal velocity. Once their terminal velocity is reached, the bubbles behave as fine-bubbles produced by a perforated diffuser.

3.4.2.1 Case study: pump mixer in sequencing batch reactor

The Blacks Ford Regional Water Reclamation Facility (BFRWRF) treats 11 ML/d (3 MGD) in four SBR tanks. The SBR tanks at the BFWRF incorporate pumped jet mixing and jet aeration. This type of mixing/aeration system relies on recirculation of the contents of the SBR tank through a manifold with a series of high velocity discharge nozzles at the bottom of the SBR tank. The system is aerated by mixing the pumped flow with atmospheric air pressurized by blowers.

There was a concern that inadequate mixing in the SBR system during un-aerated pumped mix cycles might be inhibiting the ability of the SBR to remove nitrogen. Figure 3.7 shows a photograph of the surface of one of the SBR units during a mixing cycle. The presence of an algal biomass suggests that mixing at the surface is limited and enough of a clear zone exists to allow the algal growth and even the development of an algal mat. This resulted in field tests and computational fluid dynamics investigations to evaluate the pumped and the air mixing in the SBR system at the BFRWRF.

Field tests were conducted to establish solids concentration profiles under normal operating conditions, taken at multiple locations. Mixed liquor suspended solids (MLSS) measurements were made: (1) during mixed cycles with both air and pumping operational; and (2) during pumping-only mix cycles. The measured MLSS concentration for the day of the test was approximately 2400 mg/L and the sludge volume index (SVI) was approximately 100 mL/g.

Figure 3.8 shows CFD results for the MLSS profile at various times after turning off the air: 25, 44, 53 and 66 minutes later. The profile clearly suggests that a clear layer forms at the surface and becomes more

Figure 3.7 Surface of one of the SBR Tanks in Anoxic Phase showing the presence of algae (Pretorius *et al.*, 2015).

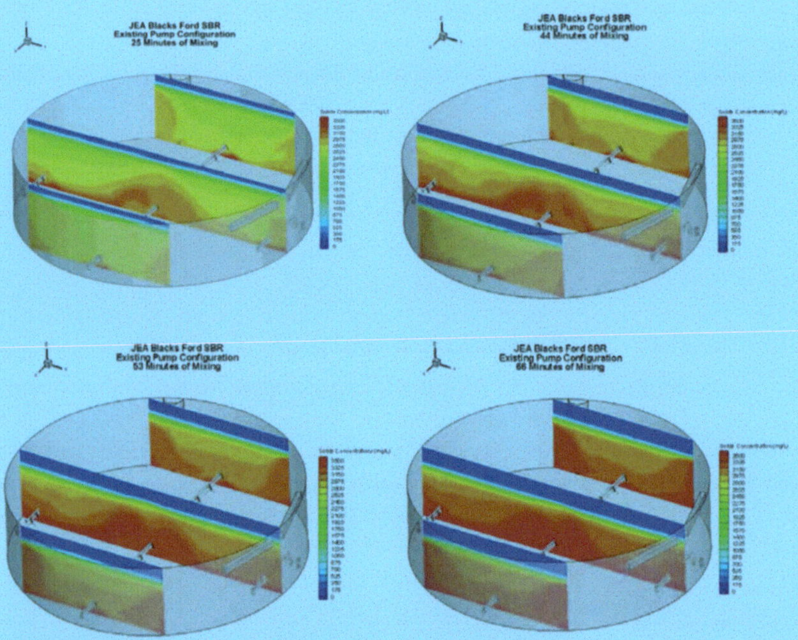

Figure 3.8 CFD calculated MLSS profiles at various times after turning the air off (Pretorius *et al.*, 2015).

prominent over time. The results are not surprising: the jet mixer is a directional mixer that is typically, and in this application, specifically, directed horizontally and downward. In the absence of air, which helps to extend its influence to the upper layers, the upper layers would experience limited impact from the jet mixers, allowing the clear zone to develop.

One way to improve mixing in a jet mixing system would be to simply increase the flow through the existing nozzles. Indeed, many existing jet mixing systems have been upgraded this way. The other factor is that of scale. The layout is such that there is a distance of approximately 12 m (39′) between the central main header and the smaller supplemental headers. This is not an exceptional distance for a mixer: aerators in oxidation ditches frequently affect conditions much further away from the input of energy. However, there are two differences between the jet mixer in this application and the surface aerator in an oxidation ditch: (1) the aerator supplies a large amount of energy, due to the limitations of surface aeration efficiency and (2) the surface aerator creates a linear flow velocity that is further enhanced by other surface aerators, so that the aerators contribute to a bulk flow pattern that helps to keep the solids in suspension. In this application, the jets do not cooperate to form a mutually reinforcing flow pattern in the tank.

Table 3.4 CFD model results for increased jet mixing power levels.

Nozzle velocity, m/s (ft/s)	2.5 (8.3)	3.0 (9.8)	3.5 (11.4)	4.0 (13.1)
Power, kW (hp)	26.0 (34.9)	43.9 (58.9)	69.7 (93.5)	104.0 (139)
Power Level, W/m^3 (hp/mil gal)	7.6 (38)	12.9 (64)	20.5 (102)	30.5 (153)
Calculated CoV	23.4%	19.1%	6.9%	5.1%

Rather, the jets on the main header and those on the supplemental headers oppose each other horizontally, potentially negating some of each other's mixing power and velocity.

When the overall pumped flow is increased, the velocity through the nozzles would increase in direct proportion to the flow. The head needed to pump the flow would increase by a square of the increased flow. Since pump power is equal to the product of head and flow (assuming efficiency does not change significantly), power to operate the mixer would increase by a cubic proportion of flow. The results of this CFD analysis are summarized in Table 3.4.

The table shows that the existing jet mix system would require relatively high power levels (20 W/m^3 or 100 hp/mil gal) to achieve a CoV of 10% or lower. As discussed above, this may be due to the distances involved and the fact that the directional mixer has less impact at the upper levels of the basin. There is also the head loss in the piping system, which would consume a lot of power, especially at higher flows, while offering no contribution to mixing.

Field test results and CFD model results confirm that the existing jet mix system cannot maintain a well-mixed solids profile throughout the depth of the existing units and that it would require almost a three-fold increase of pump power to achieve that. It also suggests that increasing the power level beyond that level would achieve only a modest reduction in CoV.

3.4.3 Air-powered mixing

Air-powered mixing, or sometimes provided by the aeration system, function by releasing air bubbles at depth, usually close to the floor, and using the bubble's upward velocity for mixing purposes. Air-powered mixing include *all* fine bubble aeration systems, even when the focus is on oxygen transfer. Note that many fine bubble systems are designed to maintain a certain minimum airflow, considered the mixing limit. Fine bubble systems can be either full floor or non-full floor. Other air-powered mixers include coarse bubble and large bubble systems.

Although mixing caused by aeration can be considered just a side effect, it is actually more than that, for it promotes a full utilization of the entire volume of the activated sludge tank and, consequently, a uniform spreading of DO, which leads to a more effective process. However, the homogeneity can be debated, as perfectly mixed conditions cannot be guaranteed in large volume aeration tanks (Gresch *et al.*, 2011a, b; Rehman *et al.*, 2014).

For full floor fine bubble aeration systems, limited data suggest that adequate mixing can be achieved at air fluxes 50% lower than the traditional guideline of 2.2 $Nm^3/m^2 \cdot h$ (0.12 $SCFM/ft^2$), and potentially lower. Depending on the mixer technology used and the geometry of the mixed zone, the ability to distribute the mixing energy throughout the water column may be a major advantage. Mixed zones with odd shapes, as well as deep or narrow zones, would be well suited for air-powered mixing.

Many air-powered diffusers are designed specifically to induce bulk flow, in particular large and coarse bubble diffusers, but also some fine bubble mixers, for example when used in spiral roll configurations. Depending on the direction(s) of the induced bulk flow and diffuser layout, plug flow conditions may be maintained or not. Air-powered mixers have the following advantages:

- Mixing energy is distributed throughout the entire water column,
- No moving parts below the liquid surface, and
- The ability to distribute mixing energy throughout the zone in a controlled manner.

The main benefit of an air mixer is that the effect of the mixing power is distributed vertically through the entire water column from the point of release to the surface. The main concern with using air mixers in unaerated zones is incidental oxygen transfer. Transfer is proportional to the surface area available for transfer, so one way to reduce transfer is to produce large bubbles, such as with coarse bubble diffusers or large bubble mixers. Further details on mixing unaerated zones using air power can be found below.

3.4.3.1 Air for unaerated zones

When using air-powered diffusers in unaerated zones, there will always be operators and engineers who recoil at the risk of oxygen transfer interfering with the anoxic or anaerobic character of the unaerated zone. For large-bubble systems, these risks have been addressed by doing side-by-side testing with non-air-powered mixers.

For coarse-bubble systems in anaerobic zones, stable biological phosphorus removal is possible. For full-floor fine-bubble systems it has been shown that biological phosphorus removal can be maintained in these systems, to improve sludge settling characteristics, by simply turning the airflow down sufficiently. For the non-full-floor fine-bubble mixer, the ability to maintain anaerobic conditions has apparently not been demonstrated, but it stands to reason that if a full-floor system can achieve it, the non-full-floor system, which potentially requires less air, could also achieve it.

The concept of intentionally aerating an anoxic zone for the purpose of promoting both nitrification and denitrification was apparently first investigated at the 91st Avenue Plant, in Phoenix, Arizona (Stensel *et al.*, 1994). In subsequent years, simultaneous nitrification–denitrification has been demonstrated as viable. In such systems, the requirement for low-level aeration while maintaining a homogeneous suspension of mixed liquor is obvious. Using the wrong mixing criteria can result in the unnecessary installation of mechanical mixing with the associated cost, as well as maintenance efforts.

When using a channel to transport mixed liquor from an aeration basin to a secondary clarifier, the channel contents are often kept in suspension using an air-powered mixer. This appears to be an ideal application for using fine-bubble aeration, along with the power levels associated with optimum flocculation. Since different researchers have found slightly different optimum power levels, designing adjustable airflow rates allowing the operators to fine tune the exact optimum power level for a given system is recommended.

A simple test to verify the effectiveness of using air for mixing unaerated zones was suggested by J.B. Neethling (personal communication, 2018): drop a sludge judge in the reactor and use the sample collected from the reactor to visually observe whether any sludge blanket formation occurs at the floor.

3.4.3.2 Hydraulic mixing: using only flow energy

In theory, it should be possible to use the energy in the flow passing through a mixed zone to keep the solids in suspension. The main challenge with relying on flow energy is that the system would require a minimum flow to function properly, and if flows drop below such a minimum, the operators would have no recourse. A system designed to maintain the sweeping velocity (and the corresponding head loss) at low flows, may struggle to pass peak flows, when the head loss would be that much higher. Nonetheless, it is an option which may be of interest in certain applications.

3.4.3.3 Case study: mixing without a mixer?

In a WRRF in Boulder, Colorado, an opportunity arose to investigate flow-induced mixing. A plan view of one of the activated sludge reactors is shown in Figure 3.9. The feed to Zone 1 consists of basin influent, MLR and RAS, which are combined in a chimney structure before entering the zone at floor level. Flow from Zone 2 to Zone 3 passes through another chimney structure. In summer two of the three reactors are in service, while in winter all reactors are put in service. On Christmas Day 2013, a historic low-flow day, the operators realized that in one zone of the plant a floating mixer was out of service and that a 2.7–3.7 m (9–12′) blanket had accumulated. The operators rushed to put the unit back in service. A later review of the data

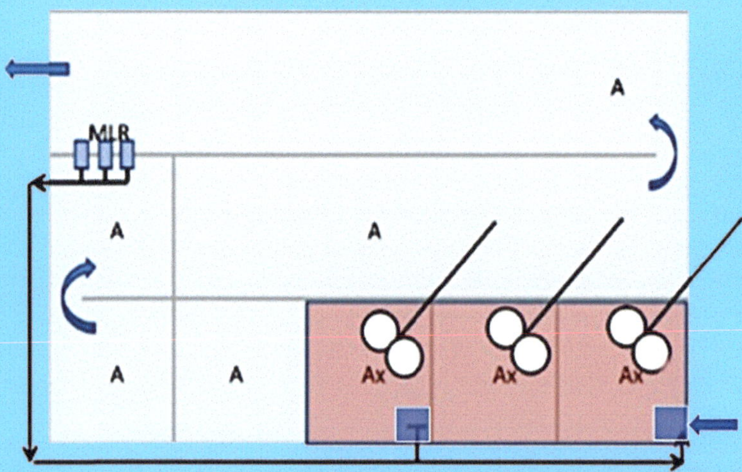

Figure 3.9 Basin layout for 75th Street Wastewater Treatment Facility in Boulder, Colorado, USA.

showed that in spite of the large mass of solids that accumulated, neither the effluent quality nor the MLSS concentration in the clarifier feed showed obvious deterioration. This prompted the operators to test the impact of taking zone mixers out of service in three different anoxic zones.

When the surface mixers were off in the three anoxic zones during the trial a blanket depth was recorded in several locations, as it happened before. The blanket level was 0.75–1.1 m in Zone 1, 0.15 m in Zone 2 and no blanket was detected in Zone 3 (Figs. 3.10 and 3.11). Flow data from the SCADA

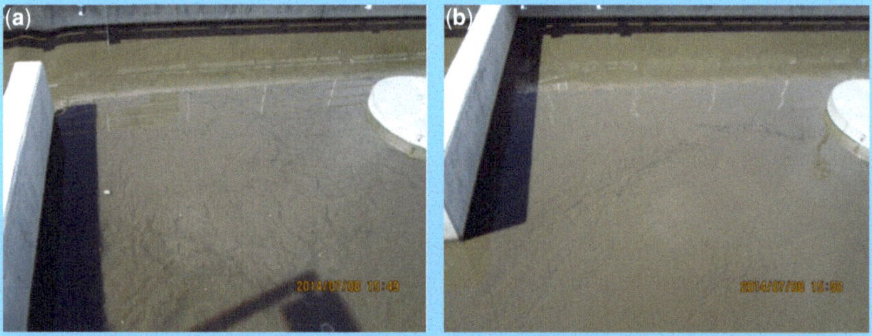

Figure 3.10 (a) Photograph of Zone 1 on 8 July 2014 (Pretorius *et al.*, 2015); a clear layer can be seen in the front left/centre with some 'rivulets' of clear liquid on the right. (b) Photograph of Zone 2 on 8 July 2014 (Pretorius *et al.*, 2015).

Figure 3.11 Photograph of Zone 2 on July 8, 2014 (Pretorius *et al.*, 2015).

system for influent and RAS flow was used to calculate the fluid velocity and head loss. These data could in turn be used to calculate the flow-induced energy that the combined feed flow imparts on the liquid, using standard fluid mechanics (Figure 3.10).

The CFD results quantified the CoV for each zone under the two flow conditions: the initial one corresponds to winter (with very low flow) and the second flow conditions corresponds to summer (higher flow) (Table 3.5).

The results suggest that under summer (high flow) conditions, Zone 2 was well mixed (CoV ~ 5%), Zone 1 was close to well mixed (CoV just over 10%), while Zone 3 was not well mixed, mostly due to a large clear zone at its surface. Under winter conditions, all the zones appear to be not well mixed (CoV > 10%). It should be noted that the chimneys divert the bulk of the flow to the floor level in Zones 1 and 3, while much of the flow is passed along the surface into Zone 2. The blanket at the floor of Zone 2 and the lack of a clear

Table 3.5 CoV for the unaerated zones with mixers out of service.

Location	CoV		Power level, mW/m^3 (hp/mil gal)	
	Winter	Summer	Winter	Summer
Zone 1	12.9%	10.6%	2.5 (0.013)	12.3 (0.059)
Zone 2	15.1%	5.7%	50 (0.25)	160 (0.83)
Zone 3	25.6%	18.8%	9.4 (0.048)	41 (0.21)

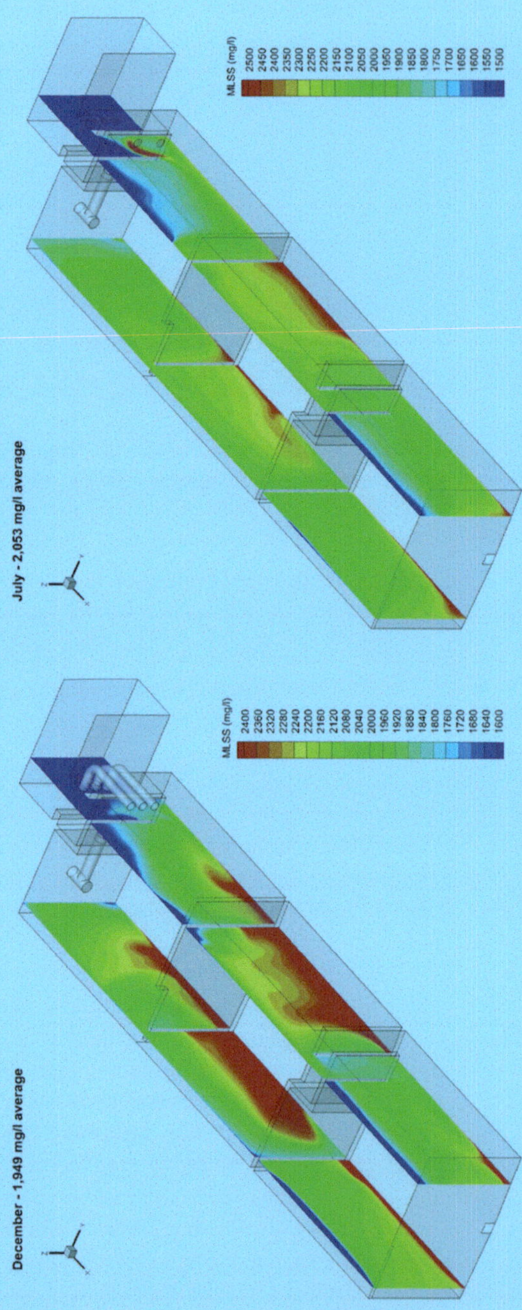

Figure 3.12 Longitudinal MLSS profile for December 25th, 2013 (left) and for July 8th, 2014 (right; Pretorius *et al.*, 2015).

zone at the surface may be related to these differences with the two other zones. The general conclusion is that using overall parameters, such as power level induced by the flow, by definition communicates limited information and that details such as the level where the flow, or the levels where portions of the flow, is introduced also matters. To include the impact of the inlet details, the design engineer may have no choice but to use CFD techniques (Figure 3.12).

The longitudinal MLSS profile from CFD analysis for Zone 2 (Figure 3.12) reveals that it is by no means completely mixed, and suggests that even at a CoV of just over 5% there is some blanket formation on the floor. Considering the solids profiles in combination with the CoV results it appears that CoV is more sensitive to the formation of a clear zone at the top than a blanket at the bottom. In terms of tracking overall conversion rate, a clear zone would have a large negative effect, whereas a sludge blanket would support conversion rates not limited by low biomass concentrations and would potentially support mixed liquor fermentation, which may be desirable. For these reasons, it would seem entirely appropriate that the CoV is more sensitive toward having a clear zone at the top. The City was interested in determining whether a modest blanket (<1 m or <3 ft) would be beneficial to overall nitrogen removal.

It is also interesting that in summer, in the second zone a CoV of just over 5% is achieved, from the flow energy only, suggesting that a power level of 0.15–0.20 W/m^3 (0.75–1.0 hp/mil gal) would be required to achieve acceptable mixing when relying solely on flow-induced power for mixing. Conceptually, one could design a system to achieve these power levels by targeting certain minimum flow velocities. One might even include gates that would allow the operators to adjust floor level openings and thus maintain minimum flow velocities and power levels over a greater range of flow conditions. It should be noted that in general one would expect a CoV calculated from CFD results to be somewhat higher than that determined by sampling, as the CFD results are equivalent to sampling every location in the basin.

As a result of this investigation, the City subsequently operated the anoxic zone mixers intermittently, each operating only 30 minutes per day, with the operation staggered. With three basins in service, each with three 4 kW mixers, the savings translated to 846 kWh per day and 32 kW in peak demand, saving the City an estimated $25,000 per year (Sigmon *et al.*, 2016).

3.5 REFERENCES

Ahnert M., Kuehn V. and Krebs P. (2010). Temperature as an alternative tracer for the determination of the mixing characteristics in wastewater treatment plants. *Water Research*, **44**, 1765–1776.

Barnard J. L., Dunlap P. and Steichen M. (2017). A comprehensive theory to understand all biological phosphorus removal observations. In: *Proceedings of WEF Nutrient*

Symposium, 12–14 June 2017, Fort Lauderdale, Florida, USA. WEF, Alexandria, VA, USA.

Braun D. and Gujer W. (2008). Reactive tracers reveal hydraulic and control instabilities in full-scale activated sludge plant. *Water Science and Technology*, **57**, 1001–1007.

Cekinski E., Giulietti M. and Seckler M. M. (2010). A new approach to characterize suspensions in stirred vessels based on computational fluid dynamics. *Brazilian Journal of Chemical Engineering*, **27**, 265–273.

Daigger G. T. (1995). Development of refined clarifier operating diagrams using updated settling characteristics database. *Water Environment Research*, **67**, 95.

Das D., Keinath T. M., Parker D. S. and Wahlberg E. J. (1993). Floc breakup in activated sludge plants. *Water Environment Research*, **65**, 138–145.

De Clercq B., Coen F., Vanderhaegen B. and Vanrolleghem P. A. (1999). Calibrating simple models for mixing and flow propagation in waste water treatment plants. *Water Science and Technology*, **39**(4), 61–69.

EPA (1989). *Respirometric methods for determination of biodegradability and biodegradation kinetics for hazardous organic pollutant compounds, EPA/600/A-92/174*. Cincinnati, Ohio, USA.

Füreder K., Svardal K., Frey W., Kroiss H. and Krampe J. (2018). Energy consumption of agitators in activated sludge tanks – actual state and optimization potential. *Water Science and Technology*, **77**, 800–808.

Gresch M., Braun D. and Gujer W. (2011a). Using reactive tracers to detect flow anomalies in water treatment reactors. *Water Research*, **45**, 1984–1994.

Gresch M., Armbruster M., Braun D. and Gujer W. (2011b). Effects of aeration patterns on the flowfield in wastewater aeration tanks. *Water Research*, **45**, 810–818.

Jiang Y., Bebee B., Mendoza A., Robinson A.K., Zhang X. and Rosso D. (2018). Energy footprint and carbon emission reduction using off-the-grid solar-powered mixing for lagoon treatment. *Journal of Environmental Management*, **205**, 125–133.

Karpinska A. and Bridgeman J. (2016). CFD-aided modelling of activated sludge systems – a critical review. *Water Research*, **88**, 861–879.

Krampe J. (2011). Full scale evaluation of diffuser ageing with clean water oxygen transfer tests. *Water Science and Technology*, **64**, 700–707.

Laurent J., Samstag R. W., Ducoste J. M., Griborio A., Nopens I., Batstone D. J., Wicks J. D., Saunders S. and Potier O. (2014). A protocol for the use of computational fluid dynamics as a supportive tool for wastewater treatment plant modelling. *Water Science and Technology*, **70**, 1575–1584.

Metcalf and Eddy (1972). *Wastewater Engineering: Collection, Treatment, Disposal*, G. Tchobanoglous (ed.), 1st edn, McGraw-Hill, New York, NY, USA.

Metcalf and Eddy (1979). *Wastewater Engineering: Treatment, Disposal, Reuse*, G. Tchobanoglous and F. L. Burton (eds), 2nd edn, McGraw-Hill, New York, NY, USA.

Metcalf and Eddy (1991). *Wastewater Engineering: Treatment, Disposal, Reuse*, G. Tchobanoglous and F. L. Burton (eds), 3rd edn, McGraw-Hill, New York, NY, USA.

Metcalf and Eddy (2003). *Wastewater Engineering: Treatment and Reuse*. G. Tchobanoglous, F. L. Burton and H. D. Stensel (eds), 4th edn, McGraw-Hill, New York, NY, USA.

Metcalf and Eddy (2014). *Wastewater Engineering: Treatment and Resource Recovery*. Chapter 3 Wastewater Flowrates and Constituent Loadings. McGraw-Hill Education.

Parker D. S., Kaufman W. J. and Jenkins D. (1971). Physical conditioning of activated sludge floc. *Journal of the Water Pollution Control Federation*, **43**, 1817–1833.

Paul E. L., Atiemo-Obeng V. A. and Kresta S. M. (2004). *Handbook of Industrial Mixing: Science and Practice*, Wiley, Hoboken, New Jersey, USA.

Pretorius C. F., Wicklein E. A., Rauch-Williams T., Samstag R. W. and Sigmon C. M. (2015). How oversized mixers became an industry standard. In: *Proceedings of the Water Environment Federation, WEFTEC 2015*, 26–30 September 2015, Chicago, IL, USA, pp. 4379–4411.

Rehman U., Maere T., Vesvikar M., Amerlinck Y. and Nopens I. (2014). Hydrodynamic–biokinetic model integration applied to a full-scale WWTP. In: *Presented at the 9th IWA World Water Congress*, Lisbon, Portugal. https://lib.ugent.be/catalog/pug01:5788243

Samstag R. W. and Wicklein E. A. (2014). A protocol for optimization of activated sludge mixing. In: *Proceedings of the 4th IWA/WEF Wastewater Treatment Modelling Seminar*. Spa, Belgium.

Samstag R. W., Wicklein E. A., Reardon R. D., Leetch R. J., Parks R. M. and Groff C. D. (2012). Field and CFD analysis of jet aeration and mixing. In: *Proceedings of the Water Environment Federation, (WEFTEC 2012)*, pp. 4113–4139.

Schauer P. and Kobylinski E. (2010). Stirred not shaken anoxic/anaerobic zone mixer sizing. In: *Presented at the Pacific Northwest Clean Water Association Annual Conference & Exhibition*, 24–27 October 2010, Bend, OR, USA.

Sharma A. K., Guildal T., Thomsen H. R. and Jacobsen B. N. (2011). Energy savings by reduced mixing in aeration tanks: results from a full scale investigation and long term implementation at Avedoere wastewater treatment plant. *Water Science and Technology*, **64**, 1089–1095.

Sigmon C. M., Pretorius C. F. and Rauch-Williams T. (2016). Plug it, cut it, drop it: improving nitrogen removal while cutting energy usage at the City of Boulder's WWTF. In: *Poster presented at WEF/IWA Nutrient Removal and Recovery Conference*, 8–12 July 2016, Denver, CO, USA.

Stamou A. (2008). Improving the hydraulic efficiency of water process tanks using CFD models. *Chemical Engineering and Processing: Process Intensification*, **47** 1179-1189.

Stensel H. D., Albertson O. E. and Hendricks P. (1994). Evaluation of anoxic-aerobic treatment at the phoenix 91st avenue plant. In: *Proceedings of the Water Environment Federation, WEFTEC Conference*, 1994.

Wahlberg E. J., Keinath T. M. and Parker D. S. (1994). Influence of activated sludge flocculation time on secondary clarification. *Water Environment Research*, **66**, 779–786.

WEF (2010). *MOP 8: Design of Municipal Wastewater Treatment Plants*, 5th edn, Water Environment Federation, Alexandria, VA, USA.

Chapter 4

Aeration control – fundamentals

Gustaf Olsson[1], Linda Åmand[2], Leiv Rieger[3] and Bengt Carlsson[4]

[1]*Lund University*
[2]*IVL Swedish Environmental Research Institute*
[3]*inCTRL Solutions Inc.*
[4]*Uppsala University*

4.1 MOTIVES FOR CONTROL

Feedback control is a powerful tool. It makes it possible to reduce the effect of disturbances and process variations. It can make good systems from poorly performing components and it can stabilize unstable systems. Biological systems make use of feedback in an extraordinary number of ways, on scales ranging from molecules to cells to organisms and further on to ecosystems. We use feedback more or less consciously every minute of our life. Information from our feet and our balancing sensors is processed to our muscles to keep us walking or standing. Oxygen levels in the bloodstream are fed back to our respiratory system to keep us going. The thermostat in every room at school or work controls the room temperature based on feedback. Driving a car requires a lot of feedback. Sensory information from our eyes, ears, and balancing system are processed in the brain and forwarded to the 'actuators', our muscles, to steer the car, keep the speed or avoid unexpected obstacles.

Editor: Diego Rosso
doi: 10.2166/9781780407845_0109

The term *control* has many meanings and often varies between communities. In this book, we define control to be the technique to get a system to behave in a desired manner despite disturbances (i.e., perturbation of a steady-state). This is typically achieved by using feedback.

A modern controller senses the operation of a system, compares it against a desired response, computes corrective actions – sometimes based on a model of the system's response to external inputs – and actuates the system to effect the desired change. This basic *feedback loop* of sensing, computation and actuation is the central concept in control.

In wastewater treatment, the motivation for all feedback control is the presence of disturbances. The load to a wastewater treatment plant is hardly ever constant. Even if that were constant, the power utility's generation portfolio is not constant, hence the relationship between the water and the power utility is not at steady state.

The flow rate, the concentrations and the composition of the influent wastewater are changing all the time. In many cases there is an order of magnitude difference between the lowest and the highest load during the day. For example, in areas where rain is a daily feature (such as the Netherlands), the peak is often more than seven times the average hydraulic load. Many disturbances are created internally, within the plant, because of the operation. Several recycle flows can cause major operational problems unless they are carefully controlled, such as the return sludge flow rate, nitrate recirculation, digester supernatants, or backwash filter flows. With all these disturbances, it is impossible to operate any plant successfully with constant settings of pumps, compressors, valves and other actuators (*actuator* is the generic term for any mechanical or electromechanical apparatus that physically implements the control action instigated by the controller's signal).

Instrumentation, control and automation (ICA) in wastewater treatment has a history of many decades (Olsson, 2012; Olsson *et al.*, 2014). At the basic level ICA is applied to keep the plant running by automatically manipulating the actuators – pumps, valves and compressors – to keep physical variables, such as flow rates, levels and pressures around the desired values. Once this is guaranteed, at a second level, the effluent quality requirements must be satisfied. One key variable is to maintain the dissolved oxygen (DO) level sufficiently high to satisfy the requirements of aerobic microorganisms. At a third level, the goal is to run the plant as economically and efficiently as possible. In particular, the DO concentration has to be limited in order to save energy and consequently limit the carbon footprint. Furthermore, an excessive DO concentration is associated with pitfalls such as unintended recirculation of oxygen in anoxic zones and excessive sludge shear.

Aeration typically represents 50–60% of process energy demand (Brandt *et al.*, 2011; WEF, 2009) but energy consumption as high as 75% has been reported (Rosso *et al.*, 2008) (See Sec. 5.1; Aeration energy vs. process energy). Thus, DO control is motivated by:

- keeping the DO concentration sufficiently elevated to satisfy the aerobic microbial requirements, both heterotrophic for carbon removal and autotrophic for ammonia oxidation;
- ensuring that the DO concentration is not excessively low, to help curbing the generation of greenhouse gases (such as N_2O);
- ensuring that the DO level is not unnecessarily high to limit the energy demand for aeration;
- preventing DO recirculation in anoxic zones.

Expressed differently, aeration control should balance between the process requirements and the need to save energy.

4.1.1 The significance of dynamics

Every operator knows that a wastewater treatment plant is hardly ever at steady state. Instead, most flow rates and concentrations are varying with time. The fact that the plant is a dynamic system also means that the result of a corrective action will take some time; it will *never* appear immediately and *delay* becomes an inevitable feature of control. Therefore, the timescales of the process changes are very important. The dynamics of a wastewater treatment system involve a wide range of timescales, from seconds to months and typical timescales can be characterized as follows:

Fast: minutes–hours

- hydraulics and flow dynamics
- oxygen transfer
- chemical precipitation
- dissolved oxygen dynamics
- solids–liquid separation

Medium fast: hours–several hours

- concentration dynamics
- nutrient removal

Slow: days–months

- biomass growth
- changes in the solids retention time
- biogas production.

The timescale influences the design of the control strategies. One way to express the control task is – for example – to supply the right amount of air, or add the correct amount of chemicals, or move the sludge to the right place to match the substrate load, *at the right time*.

Apparently, there is a wide difference between the fast and the slow timescales. Often it is possible to separate the various control actions into different time domains. This means that in the *fast* timescale the variables that change very slowly are considered constant. For example, the DO controller can act as if the biomass is constant. In the slow timescale, for example for the control of the total sludge inventory (the sludge retention time), then the DO concentration is assumed to change instantaneously.

Sometimes the sensor time lag must be taken into consideration. Collecting DO readings with a process probe takes several seconds (a laboratory probe is certainly faster but too fast and too delicate for the process). This delay is small compared to the typical time required for a change in the airflow rate to a response in the DO concentration, which is typically a fraction of an hour. A respirometer reading will take a longer time, typically half an hour. It is obvious that such a measurement can be used only for slower corrective actions, in the order of hours.

The DO concentration varies not only in time but also in space. Consider a plug-flow aeration basin. The oxygen demand in the inlet area is comparatively high because of the combination of 'hungry' organisms and high concentrations of organic material and nutrients. Towards the outlet of the aeration tank the growth reactions are reaching completion and the oxygen demand is ruled by endogenous respiration. This means that the airflow demand is much higher in the inlet area than in the outlet area. If the air supply were uniform along the aeration tank, the DO would display a typical profile, as shown below in Figure 4.3. This problem can be reduced by using step-feed control.

4.1.2 The concept of feedback

The automation must be implemented from the bottom up, starting with all the equipment that keeps the plant running. Pumps, valves and motors must operate properly, the sensors must be calibrated and maintained and the signals have to be communicated correctly to the control system.

To get a proper understanding of control we have to distinguish between *open-loop* and *closed-loop* control. The latter is commonly referred to as *feedback control.* Many processes are manipulated without any measurements made. For example:

- sludge scrapers in a settler may be operated based only on timers;
- screens are cleaned at regular intervals;
- air compressors are turned on and off based only on timers.

These control actions are examples of ***open-loop*** control. There is no measurement made. Instead the actuator is working at regular intervals, only based on timer information. In other words: after having made the control action there is no confirmation about the result of the action. For better operation, it is necessary to make measurements.

The fundamental principle of control is ***feedback*** (Figure 4.1). The process (for example, an aeration tank, a chemical dosage system, or an anaerobic reactor) is at

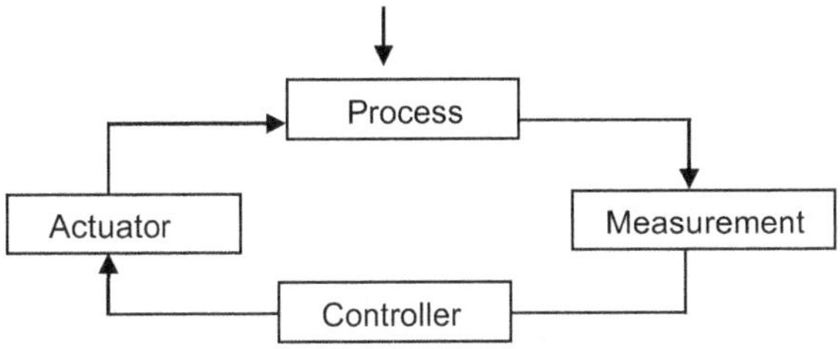

Figure 4.1 Illustration of the feedback principle. The process is influenced by external disturbances, illustrated by the arrow from above to the process.

all times subject to disturbances. Therefore, the manipulated variable can never be constant. So, the principle of feedback is simple: the control action is based on the difference between the desired and actual performance.

As already mentioned we experience feedback all the time in our lives. Any control system includes:

- measurements – a sensor system;
- signal treatment – take away all non-essential data (noise);
- a controller – the calculation of the decision, that is the control action;
- actuators – transferring the computer signal to 'muscles', to motors, valves, pumps etc.;
- a goal for the control: this may be for example 'to keep the flow rate constant', 'to maintain a constant pressure', 'to keep the DO constant', or 'to ramp up the pump speed from zero to full speed'. This goal is often expressed as the *setpoint* (or reference value).

The open-loop control example of air compressors can be converted to feedback control when the DO in the aeration tank is measured. Based on this measurement a correction of the airflow is made.

In the process industry most control loops have a constant setpoint. In such a control system disturbances will make the process output deviate from the setpoint.

No disturbances = no need for control.

If the effect of the disturbance is measured *within* the plant, such as a change in the DO concentration, a rising sludge blanket, or a varying suspended solids concentration, the measured information is *fed back* to a controller that will activate a pump, a valve, or a compressor, so that the influence on the plant behaviour is minimized. The decision taken must be implemented via an actuator, which is typically a motor, a pump, a valve or a compressor. In other words: *control is about how to automatically operate a plant or process towards a defined goal, despite disturbances* (Olsson & Newell, 1999; Olsson *et al.*, 2005).

There are other control loops where the setpoint is variable. In aeration control this happens in slave controllers in cascaded control. Then the main task of the controller is to cause the process variable to approach the varying setpoint in the best possible way. Sometimes this means as quickly as possible. At other times it means as accurately as possible.

In control engineering, **feedback control** is represented by a block diagram that describes the signals of the control system, see Figure 4.2. This kind of a simple control loop appears in almost all local control loops in wastewater treatment operation. This is aiming just to keep the plant running. Typical examples are level control in tanks for pump control, pressure control in aeration pipes, and flow rate control in various process units.

In the block diagram, we distinguish various signals:

- The *setpoint* (sometimes called the *desired value* or the *reference* value) here having the notation y_{ref}. In other words: the DO setpoint is the desired DO concentration. Sometimes the value is set manually as a constant concentration, sometimes it is variable. The latter happens when the setpoint is calculated from another controller, which is the case when DO control is based on ammonia concentration measurements. This will be discussed later in the chapter.
- The *measurement y*. Note that the sensor is not shown specifically in the diagram. The measurement value is sometimes called the *manipulated variable*, sometimes the system *output*.
- The *control error e*. In this simple case, it is just the difference between the setpoint and the measurement and the controller aims to make the error *e* as small as possible in the best possible way. For example, this may mean that the airflow (the control signal) must increase when the measured DO value is smaller than the setpoint and vice versa. The control law can be simple. In the case of Figure 4.2, the change in airflow is inversely proportional to

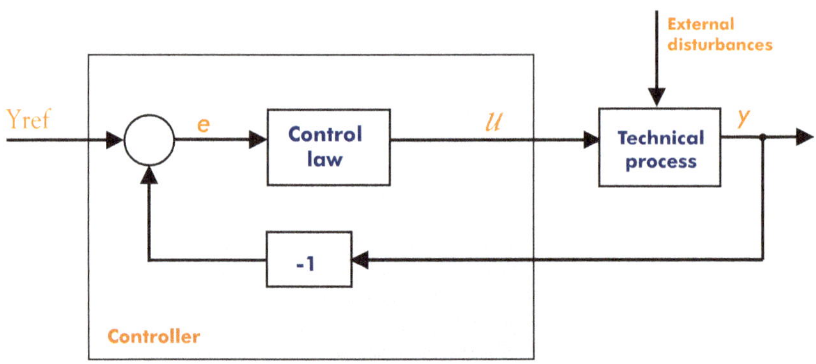

Figure 4.2 Simple feedback control structure.

the change in DO, hence the notation '-1'. It may turn on the airflow when the DO is too small and turn it off when the DO is too high. It can also be much more sophisticated: the control algorithm should maximize its information about the process dynamics to track the setpoint as well as possible.

- The *actuator* is not shown specifically in the diagram. The control law includes the algorithm that calculates the size of the control signal. The decision (control signal) must be transformed into physical changes. For a DO control system, this means a change in the airflow rate by either opening or closing air valves to control the speed of a compressor (blower). The diagram shows the control signal u, which is a combination of a computer value and an actuator.

The controller can serve several tasks:

- The most common problem in process industry applications is that the setpoint is constant and the process is subject to external disturbances, such as a load disturbance.
- Sometimes the setpoint is variable. Then the controller should track the setpoint as well as possible. This happens when an ammonia controller is connected to a DO controller in a cascade loop, see below.

Note that in both these cases the controller performs the same task: keeping the control error as small as possible. Furthermore, it is required that the controller should be robust to changes in process dynamics and avoid unnecessary changes in the control signal to reduce wear and tear of equipment.

In the design of the controller we must define what we mean by 'best'? Therefore, *controller design* and *controller tuning* become important and are discussed later in the chapter. One goal can be to keep the DO concentration constant. Another goal could be to make the DO setpoint variable so that we all the time satisfy that the ammonia concentration at the outlet is sufficiently low but not too low. We will discuss control goals in the chapter Aeration Control – Implementation.

Figure 4.2 only shows the simplest case of a single controller. In most cases there are several controllers for the system, and some of these controllers are connected to each other. We will discuss *cascade control* later in the chapter. This means that the output of one controller becomes the input (setpoint) to another controller. We talk about *control structure* when we design how controllers are connected to various measurements in the process.

Sometimes the disturbances can be measured upstream, before they hit the plant, for example the influent flow rate, or influent COD or ammonia-nitrogen concentrations. Then the information can be forwarded from such a sensor to partly compensate ahead for the disturbance. Such control action is called *feedforward*. For example, the aeration intensity can be increased or the aerated volume extended by switching on swing zones before a load increase reaches the biological treatment.

4.1.2.1 Design of DO control – concepts to consider

Some concepts to consider for the design of a DO control loop:

- DO measurements:
  - How often is it necessary (or desirable) to measure the DO?
  - Where to locate the DO sensor(s)?
  - What happens if the sensor fails?
- Measurements that may be used for feedforward control (but seldom used):
  - Influent flow rate
  - Influent COD or ammonia concentration
- Other auxiliary measurements:
  - Variables in the air production system, including the compressor states, air pressure, air valve openings, airflow rates.
- Manipulated variables:
  - The compressor speed? Add air pressure?
  - Guide vanes?
  - Air valves?
  - Air distribution along the aeration tank?
- Typical response times:
  - How often should the DO setpoint be changed?
  - From airflow rate change to DO concentration?
  - How often should the airflow rate be changed?
- How to find the best DO setpoint value?
  - Should it be fixed?
  - Can it be calculated?
  - Should it be capped? (Biofilm systems are different since they are diffusion limited)
- Controller complexity:
  - How complex does the control algorithm need to be?
  - Different strategies for low and high load?

4.1.2.2 Driving forces – demand pull and technology push

The use of ICA for the operation and management of wastewater systems will further increase in the coming years. The 'demand pull' drivers include continued population growth and urbanisation leading to increased used water load, continued increase in the complexity in the function and capability of wastewater treatment (highly energy-efficient systems and nutrient recovery systems), ever more stringent regulations, and ever stronger economic drivers. The technology development has added several 'push' factors that will make ICA even more profitable. Nowadays computational power is available virtually 'free'. There are reliable online instruments available for numerous wastewater components. There is an interesting development towards more and more 'smart' sensors, which can be placed anywhere in the processes. Actuators, mostly variable-speed drives for pumps and compressors, make control actions more flexible. Today control theory and practice can offer 'almost anything' that the water operator would need.

DO control with a constant or a variable setpoint as part of the aeration unit process operation is now a mature and proven technology. Variation in the DO setpoint is typically guided by the nitrogen removal performance monitored with ammonia sensors. This will be discussed in detail in the rest of the chapter.

A major obstacle to implement control is the lack of education in control and instrumentation. The need for education at all levels was pioneered already in 1974 by John F. Andrews (1930–2011) as he noted: 'A course in Process Dynamics and Control is commonly found in most chemical engineering curricula. We would be well advised to include a course in Dynamics and Control of Wastewater Treatment Systems in environmental engineering curricula' (Buhr *et al.*, 1974). Still there is a need today: engineers from all fields should be trained in process dynamics and modelling as well as in control theory and practice (Hug *et al.*, 2009).

4.1.2.3 Some historical notes – the early years of aeration control

As early as in the 1930s the control operation of activated sludge plants attracted attention. Palmer (1939) indicated several variables essential for good operation. These were: quantity of air used, aeration period, concentration of MLSS (mixed liquor suspended solids), settling period, rate of sludge return and sludge condition as indicated by a determination of its oxygen consumption. A couple of years earlier, Haseltine (1937) had concluded that the first operating requirement of the activated sludge process was the maintenance of DO in the aeration tanks. Heukelekian (1934) had proposed that the minimum DO concentration should be >0.5–1 mg/l, while Collier (1939) suggested that 2 mg/l should be the minimum. This was confirmed by practical experience from the North Side Works in Chicago, IL, USA. However, the use of an apparatus for quickly determining the DO demand did not gain favour in the late 1930s. Many operators were convinced 'that the plant itself is its own best odometer' (Pearse, 1941).

Process control received more attention in the 1960s as being 'no less applicable in theory to sewage treatment as it is to any other microbiological process' (Jones *et al.*, 1969). In practice, however, it was considered more difficult, since 'both the quantity and the quality of the incoming sewage are variable and it is essential that the process operates continuously and efficiently at all times of the year if receiving waters are to be protected' (ibid.). The authors also stressed that to save energy the (DO) 'concentration should not be maintained at a level any higher than necessary'. We may throw in an additional wrench by noting the unknown composition of the wastewater, which adds uncertainty to the variable concentration.

Thus *feedback control requires measurements.* Instrumentation was still the weak link in the 1970s. Some progress had been made to develop DO sensors, COD analysers and TSS meters, and these instruments were tested in some pioneering control efforts. As early as the 1950s a semi-continuous colorimeter in conjunction with the Winkler method was being used to measure DO concentration (Briggs *et al.*, 1954). Later this sensor was replaced by the dropping-mercury electrode (Briggs *et al.*, 1957). By the early 1970s the use of online DO sensors was well established in many wastewater treatment plants (Briggs, 1973; Briggs *et al.*, 1967; Brouzes, 1969; Roesler, 1974). Direct digital control of DO was performed in Palo Alto, California, USA (Petersack & Stepner, 1973), in Bridgeport, Connecticut, USA (Dobbins, 1974) and in Stockholm, Sweden (Olsson & Hansson, 1976) and very soon significant savings of air could be reported. For example, Guarino and Radziul (1977) reported a 22% reduction in air demand per unit of BOD removed. In general, the *laissez faire* practice of not controlling DO in aerated processes is the recipe for energy wastage (at best) and process failure (at worst).

Still, in the 1970s (and even later) instrumentation reliability was called into question. The control of biological systems was a matter of art rather than science. Many systems were controlled by rules-of-thumb. The fundamental kinetics upon which automation systems must be based was just becoming apparent. This naturally brought up the questions not only of 'detectors' or instruments, but also of control methods, operating equipment and computer programs. There were important developments in computers, and real-time computing for online control was becoming affordable.

The air distribution along the reactor in a plug-flow activated sludge basin could usually be controlled only manually. This was seldom done since a change of one valve opening along the reactor immediately resulted in a new airflow distribution that was not necessarily the desired one. The aeration rate in the inlet region was often insufficient to satisfy the DO requirement, while near the reactor outlet it was more than what was required (Figure 4.3).

Tapered aeration, achieved by different air valve settings along the aeration tank, was applied at the Salinas, California, USA plant from 1930 to 1933, applying 55–70% of the total air in the first half of the aeration tanks (Haseltine, 1937). Torpey (1948) found by experience that the DO profile levelled off at around 75% of the reactor length, indicating that the oxygen uptake reaction had gone to completion. Without any control, the profile was pushed towards the outlet of the aeration tank at high organic loads, while it receded towards the inlet at low loads. The profile behaviour was interpreted in Olsson and Andrews (1978) and it was demonstrated that there was not one position in the tank that could represent the DO concentration to control the airflow. Instead the *slope* of the DO concentration towards the outlet indicated whether the reaction had gone to completion. Various approaches

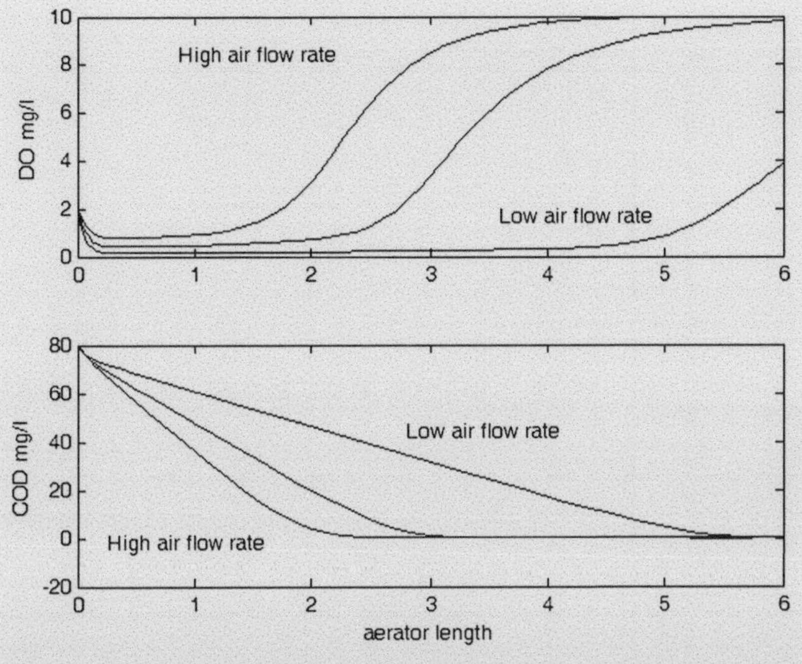

Figure 4.3 A typical DO profile in a plug-flow reactor for carbon removal with uniform airflow (diffuser) distribution. The upper figure shows the DO profile (mg/l) for different airflows. The lower figure depicts the corresponding carbon concentration (mg COD/l) decrease along the plug flow reactor (from Olsson, 2012).

were made to use two DO sensors, located along the aeration tank towards the outlet, to find the most efficient airflow (Gillblad & Olsson, 1977). Today it is well established that the problem of DO profiles can be addressed better by providing independent air supply systems to different aeration zones. Still the DO profile analysis provided insight into the DO dynamics, since it is a direct mirror of the oxygen uptake.

4.2 PROCESS DYNAMICS

The behaviour of a dynamical system changes over time. It is important to note that the response (i.e., the output) of the system depends not only on the present control signal but also on the past history of the system. It is crucial to understand the dynamics of the system for the controllers to be able to handle these changes. The slow growth of the nitrogen-removing bacteria, the limited impact of the DO concentration on the nitrification rate under certain conditions, as well as the non-linear dynamics in the air supply system are the most important messages from this chapter and will be described in more detail below.

4.2.1 Dynamics of the aeration supply system

The aeration supply equipment with blowers, pipes and valves, the sensors and the oxygen transfer from air to water has dynamics. Examples of dynamics are lag and time delays creating a delayed response to a change in a control signal but there are also non-linear relationships such as saturations.

The time for a change in the actual airflow rate from a change in the airflow set-point has been modelled to approximately 4–5 minutes by Rieger *et al.* (2006). This response time includes the dynamics of the airflow control loops, the rise time of bubbles and the time delays in the air supply system. If also the transfer of DO from air to liquid phase is included, the response time is around 15–30 minutes. This means it will take typically half an hour to reach a new DO equilibrium after a step change in the DO setpoint.

There are several non-linearities in the DO control process. Firstly, the airflow valve may be non-linear. This means that a change in the valve opening by 10% will cause a different airflow change when the valve opening is small compared to when the valve is almost fully open. Often the relative change in airflow rate is reduced when the valve approaches 100% opening. Secondly, the oxygen transfer rate from the air bubbles to the water liquid is non-linear with a reduced transfer rate at higher airflow rates, which is related to bubble dynamics. Thirdly, there is a saturation of the DO concentration in water. This saturation depends on, among other things, the dissolved solids concentration, water temperature, and air pressure. The DO saturation concentration at sea level and 20°C is 9.08 mg/l.

How do the response times and non-linearities impact on aeration control design? The response times give us information on how to tune the controllers, which is discussed in Sec. 4.3. The non-linearities mentioned above can all be described as saturations with decreased efficiency at higher valve openings, airflow rates or DO concentrations. These may call for limits in the controllers, as further discussed in Chapter 5. In this section, other constraints that impact on controller design are also mentioned, such as limitations in airflow capacity, another example of non-linear behaviour.

4.2.2 Dynamics of microbial processes in nitrogen removal

Biological wastewater treatment is carried out by various groups of microorganisms engaged in different treatment steps. Which organisms are active depends on the environmental conditions. Typically, these conditions are defined by dedicated reactor designs with aeration as one of the key drivers. The objective for aeration control at a Water Resource Recovery Facility (here we will use the WRRF terminology instead of the traditional term Wastewater Treatment Plant, WWTP) is to provide sufficient oxygen to aerobic processes and limit oxygen transfer for anaerobic (no oxygen, nitrite or nitrate present) or anoxic (nitrite and nitrate present) processes. Real-time control enables the plant to react to disturbances caused by influent load variations, temperature, etc. in a dynamic fashion and in real-time providing optimized conditions for the microorganisms required for one or the other process.

4.2.2.1 Advanced reading – microorganisms in the bioreactor

The cell synthesis in microorganisms requires, among other things, energy and a carbon source. Energy-producing reactions involve electrons changing place during an oxidation–reduction reaction. Microorganisms can produce energy and obtain carbon for cell growth from a range of chemical compounds. Table 4.1 summarizes important distinctions between microorganisms present in the biological treatment steps in WRRFs. The table indicates that air supply and aeration control are required for COD removal, nitritation, and nitrification.

In the aeration tank, carbonaceous material is removed by aerobic heterotrophic bacteria, which use carbon, measured as COD or BOD, as a carbon source and as an electron donor in the respiration, while oxygen is an electron acceptor. There are many heterotrophic aerobic microorganisms, constituting both bacteria and archaea.

Nitrogen removal is a two-step process: nitrification and denitrification. Nitrification is also a two step-process, where ammonia-oxidizing bacteria

Table 4.1 Sources of energy and carbon for different organisms working in a biological treatment process for carbon and nitrogen removal.

Biological Process	Type of Organism	O_2 (Y/N)	Carbon Source	Energy Reactant	Energy Reactant	Reaction Products
				Oxidation	*Reduction*	
COD-removal	Heterotrophs	Y	COD	COD	O_2	CO_2, H_2O
Nitritation	Ammonia-oxidizing bacteria (AOB)	Y	CO_2	NH_4^+	O_2	NO_2^-, H_2O
Nitrification	Nitrite-oxidizing bacteria (NOB)	Y	CO_2	NO_2^-	O_2	NO_3^-
Denitrification	Heterotrophs, facultative anaerobic	N	COD	COD	NO_3^-	NO_2^-
Denitrification	Heterotrophs, facultative anaerobic	N	COD	COD	NO_2^-	N_2 (g)

(AOB) are responsible for oxidizing ammonia to nitrite and nitrite-oxidizing bacteria (NOB) oxidize nitrite to nitrate. AOB and NOB are autotrophs, meaning they do not require an external carbon source since they use carbon dioxide for cell synthesis. Denitrification on the other hand is performed by a large group of microorganisms that can use oxygen as an electron acceptor – but do not have to. This means that they are facultative anaerobic. Nitrate can serve as an electron acceptor if oxygen is not available. This is referred to as an anoxic environment. However, oxygen is the most energy-efficient electron acceptor of all. Avoiding oxygen in compartments meant for denitrification is therefore important.

Nitrifying bacteria require oxygen, but they also require sufficient amounts of ammonia in the wastewater, and other nutrients needed for growth and sufficiently long aerobic solids retention time (SRT). The amount of ammonia in the influent together with the DO and the SRT will determine the amount of nitrifying biomass. The aerobic SRT in nitrifying plants is often above 10 days. Hence, the nitrification capacity changes slowly, aeration control or not.

Nutrients or energy sources can be limiting to microbial growth. This limitation is commonly modelled with a Monod function. In a Monod function the specific growth rate of microorganisms is reduced when the limiting substrate is decreased. In aeration control this kind of dynamics play an important role during nitrogen removal, since growth of nitrifiers requires both oxygen but also ammonia to proceed at maximum capacity and sludge retention time (SRT) that is high. An example of a Monod relationship is found in Figure 4.4.

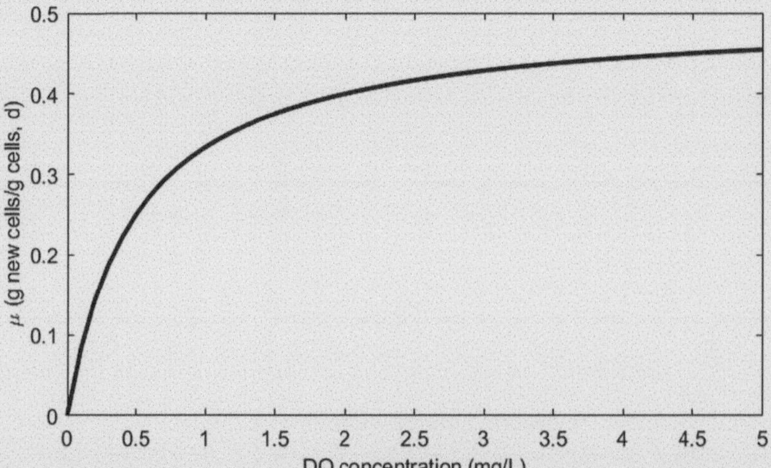

Figure 4.4 Example of a Monod curve: growth rate of an aerobic microorganism (μ) as a function of DO concentration. Supplying DO above 1.5 to 2 mg/l has only a limited impact on microbial growth, while insufficient oxygen levels (i.e., <0.5 mg/l) would limit growth.

Control *authority* is the capability of a controller to reach a desired setpoint over a range of conditions. For example, during a high-load period the air supply capacity may be insufficient. This means that the desired DO setpoint cannot be reached. There may be a similar limitation for low loads. To maintain mixing requires a certain amount of airflow. As a result, the DO concentration may be higher than the setpoint and the aeration control is limited towards low airflows.

The control authority to keep the effluent ammonia within desired limits is one-sided in the short-term timescale. The effluent ammonia can be increased by lowering the DO concentration, but during peak load the control authority is commonly lost. Increasing the DO concentration has no effect since there is a kinetic limitation of the nitrification capacity, as described by the Monod kinetics (Figure 4.4). Increasing DO above, say, 2 mg/l in the aeration tank has only a little effect on the effluent ammonia concentration, since the nitrifying biomass by then is already working at full speed.

The generation of greenhouse gases is gaining increasing attention. In the biological nitrogen removal process, nitrous oxide (N_2O) is formed through several different pathways. It is a severe greenhouse gas with a climate impact about 300 times higher than that of CO_2, and emissions should therefore be avoided. Much research into the production of nitrous oxide during nitrogen removal has generated new knowledge during the last decade. The dominating theory is that most of the nitrous oxide is produced in the nitrification reactors through autotrophic denitrification, and important risk factors are low but non-zero DO concentrations, and high nitrite concentrations. There is also a potential N_2O release due to abrupt DO changes that happen in intermittent operations (Kampschreur *et al.*, 2009, Lang *et al.*, 2016, Lindblom *et al.*, 2016). Further, the N_2O emission in the first aerated compartment is a combination of the emission from the production in the aerated tank plus the stripping of the N_2O produced in the anoxic tank and not emitted there (Townsend-Small *et al.*, 2011).

4.3 CONTROL STRUCTURES AND ALGORITHMS

Creating a control system begins with the *structure of the control*, explained in the first subsection. Having a structure, the *control algorithms* can then be defined. Finally, the parameters of the controllers must be determined. This is called *controller tuning*.

4.3.1 Control structures

Choosing a control structure means pairing the manipulated variables of the system with suitable measured variables. When feedback control is in place the measured variables will – thanks to the feedback loop – change depending on the value of the measured variable. It is the measured variable that the operator wishes to control towards a reference value, despite disturbances in the system. The simple feedback loop (Figure 4.2) is the basis for most of the controllers in a typical

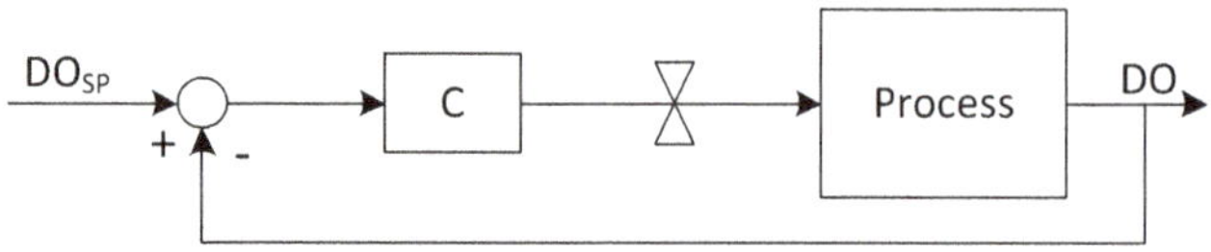

Figure 4.5 Standard feedback DO control.

WRRF. A comparison between studies in the literature comparing different control structures in aeration control is found in Åmand *et al.* (2013).

A manipulated variable for control of the aeration intensity at a WRRF is either the valve position or the compressor speed – if using diffused aeration. There are several measured variables at hand to pair with these manipulated variables. The most commonly used are airflow rate, DO concentration and effluent ammonia concentration. These three variables are commonly used in a control structure called *cascade control*, a modified version of the classical feedback loop in Figure 4.2.

If no airflow measurement is available, the DO measurement is used in a classical feedback loop to control the valve opening, as shown in Figure 4.5. In the coming figures, C is short for controller.

4.3.1.1 Cascade control

In cascade control, there are several measured variables, but only one manipulated variable, illustrated in Figure 4.6. The DO controller is called the *master controller* and receives the DO concentration measurement value and calculates the airflow setpoint to the airflow controller that serves as the *slave controller*. It compares the measured airflow rate with the setpoint and will manipulate the valve position so that the DO concentration approaches its setpoint.

The idea behind cascade control is to control a system with fast dynamics (e.g., airflow control) with a fast inner controller and controlling slower dynamics (e.g., DO dynamics) with a slower outer controller. In this way, the inner slave controller can compensate for disturbances or non-linearities in the inner control loop, leaving the system easier to control for the slower master control.

Another important reason for cascade control is that the effect of nonlinear valve characteristics is reduced. By measuring the airflow, we do not need to worry about the valve characteristics but can concentrate on the result of the air

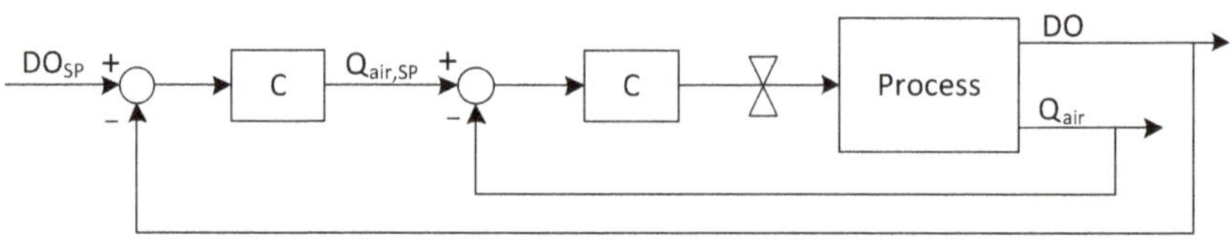

Figure 4.6 The standard cascade controller configuration of a DO control system.

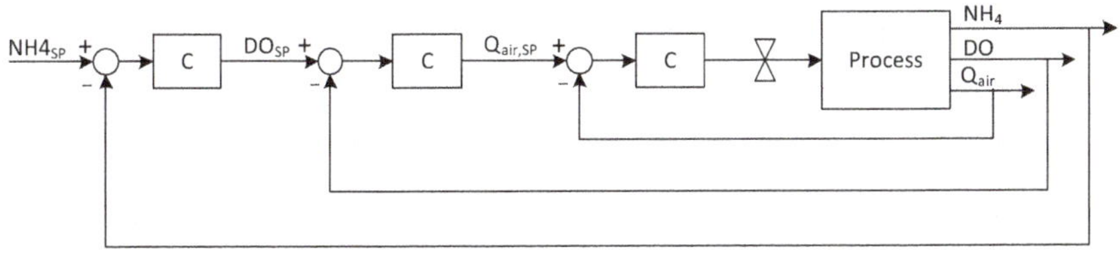

Figure 4.7 Ammonium-based DO control, using three controllers in cascade.

valve change. In this way, cascade control simplifies controller tuning and the controller commissioning (going from manual to automatic control). *Cascade control must always have a slower master controller than the slave controller.*

It is possible to have more than one slave controller. Such an example is given in Figure 4.7, where an ammonia controller is added to the control structure of Figure 4.6. Ammonia dynamics is much slower than DO dynamics, which makes cascade control a viable control structure. Now the DO setpoint is no longer constant: it is calculated by the ammonia controller. The effluent ammonia setpoint is given as a constant. If the measured ammonia concentration is higher than its setpoint, then the DO setpoint is increased.

Note that one ammonia controller can be used to provide setpoints to several DO controllers along the same treatment line. Each set of DO and airflow controllers work on each separately controlled airflow valve, while an ammonia controller typically works on each treatment line at the plant.

Feedback controllers solve most control problems at WRRFs or in other process industries. However, feedback control has one major drawback: a disturbance must make a measurable impact on the system before the controller can act. Therefore *feedforward* control is sometimes used as a complement to feedback control.

4.3.1.2 Feedforward control

Feedforward control can act on a disturbance before it affects the process to be controlled, and thereby reducing its impact. The disturbance is then measured before it hits the process. For example, the influent flow rate or the influent ammonia concentration can be measured. The feedforward controller should act before the impact of the disturbance can be seen in the output. This means that the feedforward controller needs a model of the system that is impacted by the disturbances. The goal of the feedforward controller is to take corrective action (for example increasing the airflow rate) so that the result of the disturbance will not be noted. Ideally, if the model were perfect, no disturbance would be observed in the process. However, no model is perfect. The controller is an open-loop controller, so it cannot check if the compensation has been corrected. It is not using information of the output signal: this means that there is a risk for either over-compensation (aerating too much) or under-compensation.

Therefore, a feedforward controller should be combined with a feedback controller. The feedback controller measures how much the output (such as the DO concentration) has been influenced by the disturbance, and a final correction can be made. The combination of a feedforward and a feedback controller is generally a very attractive strategy, combining a fast reaction to a disturbance with a slower but robust feedback.

An example of feedforward aeration control is found in Figure 4.8, where influent ammonia and flow-rate measurements are used as feedforward signals. We can say that the feedforward controller is used to increase the bandwidth of the control, while the feedback controller will maintain the accuracy of the control. A higher bandwidth means that the control system will react much faster to correct for a disturbance.

Feedforward control can be useful when a plant must deal with load peaks. For example, a rapid increase in the ammonia load will cause elevated ammonia effluent concentration. A feedback controller may simply be too slow to dampen the influence of the peak disturbance. The information (ammonia measurement) to the controller is coming too late. This is a problem if the peaks are causing violation of the effluent permit. For example, in Germany or Austria the effluent limits are never to be exceeded. The goal of the feedback controller is to limit nitrification to save energy and at the same time improve the conditions for denitrification. If the feedback controller is tuned for this purpose it may be too slow to act on a rapid increase in the influent ammonia. However, a feedforward controller will get an early warning by measuring the ammonia peak in the influent. One strategy can be to increase the airflow rate early on. Another strategy can be to increase the aerated volume to increase the mass of active nitrifiers. Such a change will, however, take a long time. This will take care of most of the impact of the ammonia peak. Then the feedback controller will make the final correction by measuring the effluent ammonia concentration. However, to increase the aeration intensity above say 2 mg/l of DO (in all sections of the plant) has only a limited impact.

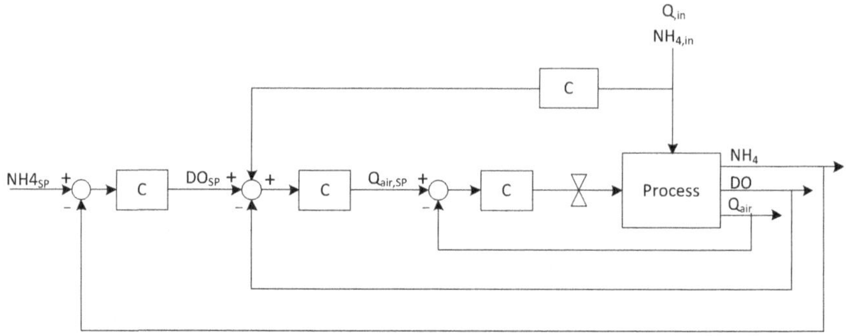

Figure 4.8 The control structure of DO control using feedforward information.

4.3.1.3 Advanced reading – multiple-input–multiple-output controllers

In control theory, a division is made between single-input–single-output (SISO) controllers and multiple-input–multiple-output (MIMO) controllers. 'Input' and 'output' refer to signals going in to and out from the controller. All the three controllers in Figure 4.7 are SISO controllers. SISO controllers are more easily tuned than MIMO controllers. In aeration control, SISO control is most often used, especially for airflow and DO control. MIMO controllers are better at taking care of interactions between process variables than SISO controllers. However, in practical applications of wastewater treatment operation there are very few examples where MIMO control is necessary. There are few processes that are truly multivariable in the sense that a multivariable controller is necessary. In most cases there are only insignificant cross-couplings between several inputs and outputs. Therefore, most process parts can be favourably decoupled and thus controlled by SISO controllers. Ingildsen (2002) discusses the pros and cons of MIMO controllers in wastewater treatment systems. The work by Halvarsson (2010) presents a deep insight into MIMO controllers.

4.3.2 Control algorithms

Control theory has had a truly extraordinary development during the last four decades. Therefore, it is quite remarkable that the control theory that was available in the 1970s can still solve a majority of the process control problems in wastewater treatment (Olsson & Newell, 1999). The control of wastewater treatment systems is certainly not limited by lack of control methods. Rather, the challenge is to achieve a comprehensive understanding of the process and its limitations so that a feasible control objective can be formulated. Furthermore, the control authority of the actuators, the reliability of and the relevant information extracted from the sensor data, the data management and monitoring strategies must be carefully considered.

The control algorithm is what happens inside the controller – how the information in the input signal is transferred to a control action. We present here an overview of the most relevant control algorithms for aeration control.

4.3.2.1 On–off control

The simplest type of control algorithm is the *on–off* controller: for example, turn on aeration if DO concentration drops below a specified limit, and turn it off when it exceeds a specified limit. The two limits do not necessarily have to be at the same concentrations. This kind of control is common in our daily life, for example temperature control in an air-conditioned room.

On–off control is the simplest type of rule based control, where an *if … then …* statement is used to decide how the controller should act. There are more sophisticated versions of rule-based control. For example, the DO concentration can be changed in steps depending on the effluent ammonia concentration. Another useful way to use rule-based control is by switching on or off aerated zones according to a specified criterion. On–off control can also be used in sequencing batch reactors to switch between nitrification and denitrification phases.

4.3.2.2 The proportional-integral-derivative (PID) control

By far the most commonly used control algorithm in controllers around the world, especially within the process industry, is the PI(D) controller. This controller was historically developed for manufacturing control and vessel autopiloting, but is applied to many processes nowadays, including aeration blowers. Most often the controller is referred to as PID, but we are purposely labelling it PI (D) because of its characteristics in our industry, seldom acting as a derivative controller (i.e., the 'D' in PID). Based on a survey of over eleven thousand controllers in the refining, chemicals and pulp and paper industries, 97% of regulatory controllers utilize PID feedback (Desborough & Miller, 2002). There are literally hundreds of textbooks on PID controllers. Hence, we are not about to write the n-th one. A little booklet (*PID Loop Tuning Pocket Guide*, 2017) summarizes the properties of the controllers, while the textbook by Åström and Murray (2016) gives a deep insight into control. Both are available online.

The PID controller calculates the control signal based on the control error, that is the difference between the setpoint (reference value) and the measured value of the key variable, such as DO. The proportional part (P) of the controller amplifies the control error with a proportional gain (K). The integral part (I) integrates the control error. Integration means to sum up something. The integral action in a controller has the useful effect that the controller will work until the control error is zero, even if the error is very small.

The derivative part of the PID controller (D) is proportional to the derivative of the control error. The derivative is nothing else than a simple calculation of the future trend. However, using the D-part of the PID controller always requires more caution. If the measurement signal is noisy, a straightforward use of the derivative will result in a very noisy signal. Therefore, filtering of the signal is required before the derivative is calculated. More details are given in the references above. For the continued discussion, we limit ourselves to the PI controllers, since they are the dominating controllers in most process industries, including wastewater treatment.

The control error is calculated as:

$$e(t) = y_{ref}(t) - y(t) \tag{4.1}$$

where $y_{ref}(t)$ is the reference value (setpoint) at time t and $y(t)$ the measured value at time t.

Mathematically, the (ideal) PID controller is defined as

$$u(t) = K\left(e(t) + \frac{1}{T_i}\int e(t)dt + T_d\frac{\mathrm{d}e(t)}{dt}\right) \tag{4.2}$$

where K is the (proportional gain). T_i is called the integral time and T_d the derivative time.

The reason that the PI(D) controllers are so successful in these applications is that most of the processes can be described by low-order dynamics. Some non-linearities (such as Monod-type ones) can be described as 'smooth' non-linearities, that is the systems still behave linearly for small variations around the operating point. Others – such as valve behaviour – can be compensated by cascade control, using simple proportional (P) or PI controllers. Highly non-linear dynamics – such as exothermic reactions – do not appear in wastewater treatment systems.

A wastewater treatment system contains both fast and slow processes, with timescales from seconds (air and liquid flow rates), to hours (concentration changes), to days and months (microbial communities). This makes it possible to decouple the plant in slow and fast modes. Fast control, for example for aeration, can be performed by assuming that the biological properties do not change within hours. Similarly, slow controls, such as waste sludge control, can be performed where fast phenomena like oxygen transfer can be considered instantaneous. This will be further illustrated in Chapter 5.

4.3.2.3 Advanced reading – model-based control

We need a model as a basis for controller design. The reason that the PID controller is successful in most applications in the process industry is that the process models for control can be quite simple. Analysis methods can be combined with an empirical approach to the problem to design controllers with an acceptable level of performance. During recent decades more powerful control techniques have been developed and used in various industrial sectors to increase their productivity.

A typical example is that of *model predictive control* (MPC). An interesting aspect of MPC is that it was born more than 30 years ago and developed in the industrial world, based on a pragmatic approach to what optimal industrial control could be. It only began to arouse the interest of the scientific community much later. The basic idea is to predict the behaviour of the process at a certain time horizon *T*. Given the desired performance, we will know where the process should be after the time *T* (Figure 4.9).

The controller now can calculate how much to manipulate the system so that the predicted value reaches the desired value. An MPC controller can perform much better than a PID controller in cases where there are hard constraints on the state variables or the control signal, nonlinearities or

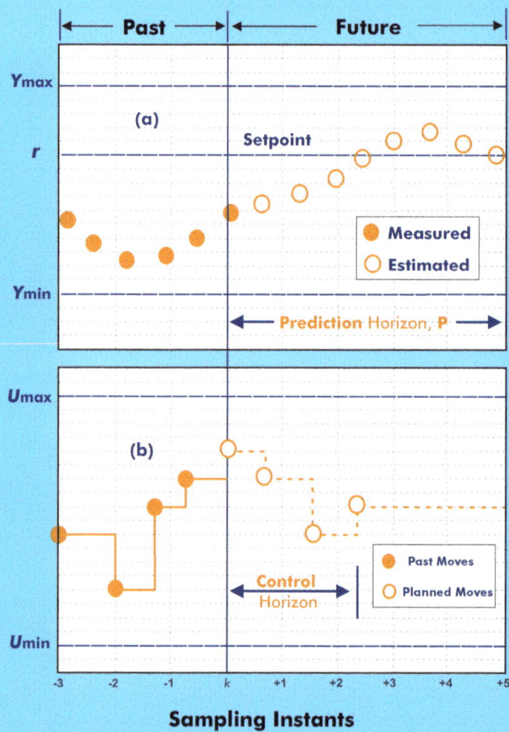

Figure 4.9 A model predictive controller can predict the behaviour of the process. A model of the process estimates the future behaviour of the variable. In the figure, the variable has been measured up until the sampling instant k. The target of the variable is the setpoint and the estimates are based on the control signal until time k. The controller now calculates a control signal that will bring the process closer to the setpoint. At the next sampling time $k+1$ the procedure is repeated another time step.

significant time delays. As an example, how much can we improve the operation of a process if we can predict the influent load 1, 2, or 12 hours ahead?

4.3.2.4 Advanced reading – gain scheduling

A DO control system behaves differently at different loads. For example, the oxygen transfer (K_La) is not linearly depending on the airflow rate (see Chapter 2.1). At low airflow rates, the oxygen transfer is relatively sensitive to an airflow rate change, while at high airflow rates it is less sensitive. Similarly, as mentioned in Sec. 4.2, the airflow valves are nonlinear.

Because of the nonlinearities the best PI(D) regulator tuning is different at low and high loads. If the controller is tuned for low loads it means that the process is more sensitive to changes, indicating that the controller gain must be smaller. With this gain the controller will react too slowly at high loads. Conversely, having tuned the controller for high loads it may be too sensitive at low loads and create oscillations of the process variables. Normally the controller is tuned so that it would behave satisfactorily, if not perfectly, at all loads.

With gain scheduling, the tuning of the PI(D) controller will be changing depending on the load. These values are calculated in advance. The sensors of primarily the airflow and valve positions can signal the current values to the controller to change its tuning parameters. Such a feature will enhance the performance on the PI(D) controller.

Obviously, the control computer must determine when the process has entered a new operating range. It then updates the controller with a predetermined set of tuning parameters designed to optimize the closed-loop performance in that range. Unfortunately, gain scheduling also means a lot of work. The control engineer implementing the gain schedule must first determine how the full span of the process variable should be partitioned into distinct operating ranges that adequately represent all the possible variations in the process's behaviour.

The implementer would then have to operate the process within each range and tune the controller for optimal closed-loop performance each time. Then s/he would load the resulting sets of tuning parameters into the gain schedule to be retrieved by the controller whenever the process variable enters the operating range that corresponds to each set. For the DO control that would require the implementer to tune the controller at a certain load, maintain the load there, tune the controller, and record the resulting tuning parameters in the gain schedule. Then this must be repeated until the entire range loads have been tuned.

However, the good news is that a modern PID controller can solve the gain scheduling problem readily. It boils down to making a tuning at each range, for example using autotuning. Therefore, the user does not need to worry about implementing the actual scheduling.

4.3.3 Defining the control goal

As already discussed, specifying the goal of the control system is a key challenge. It must include specifications like:

- attenuation of load disturbances,
- sensitivity to measurement noise,
- robustness to model uncertainty,
- setpoint following.

The last item is not common in aeration control. In general, a trade-off needs to be done since, for example, a fast attenuation of disturbances may imply that the system reacts too sensitively to measurement noise. There are fundamental limitations in all control design. Typical sources of limitations are:

- the dynamics of the controlled process,
- control authority (compare the ammonium feedback in Sec. 4.3.1),
- nonlinearities,
- disturbances.

A consequence of nonlinearities is for example that the controller tuning may be quite different for different loads. Process uncertainty is another challenge. A biological system may change character if the composition of the feed changes. New organisms will appear that require a different kind of control action, for example filamentous growth.

4.3.4 Controller tuning

Here we will describe some key ideas to tune the two parameters in a PI controller.

Any car driver will learn what 'tuning' means. We move the steering wheel to follow a winding road. If the 'gain' (K in the controller) is too large, then we will turn the steering wheel too fast to keep the car in the middle of the lane all the time. Such a 'wild beginners' control can easily result in overcompensation (instability), which means a high risk of an accident. On the other hand, if the driver compensates the position too slowly (a small K) then the performance of the driving is too poor. A good driver knows how much the steering wheel must be moved to compensate for a certain error in the position in the lane.

A key to success in a control system is to adjust the control parameters to obtain a desired behaviour – the tuning part. It is worth stressing that there exist no 'default values' or 'plug-and-play' for the parameters in a controller. Those need instead to be calibrated/tuned according to the dynamics of the process to be controlled in conjunction with the control objective. The tuning could be based on process models as described in numerous textbooks on automatic control, see for example Åström and Murray (2016). For tuning a PI(D) controller, dedicated empirical methods have been developed (Åström & Hägglund 1995, 2006). The basic idea is to perform a simple experiment and extract some characteristics of the process dynamics from the experiment and determine the controller parameters from these, basically applying rules-of thumb. The first empirical tuning rules were developed over seven decades ago by Ziegler and Nichols (1942). It should be mentioned that the Ziegler–Nichols rules typically give too 'aggressive' a tuning for wastewater treatment operations.

It is worth stressing that empirical tuning may only give a 'ball-park' of control parameter values. Hence, it may be necessary to perform manual tunings. The following rules are useful:

- Increasing the proportional gain (K) increases the speed of response, but decreases stability and increases the noise sensitivity.

- Decreasing the integration time (T_i) decreases the control error more rapidly, but also decreases stability.

4.3.4.1 Advanced reading – lambda tuning

Lambda tuning is simple and intuitive to use, and is common in the process industry. We regard lambda tuning as a typical default choice for tuning a controller for aeration control. As with many other methods, it is based on making a step response on the open-loop system and measuring a few characteristic parameters. An important issue is what units the control uses internally. Even though data from the step response may be presented in engineering units it is common that the controller internally works in units of percentage span.

The lambda tuning can be summarized in four steps (note that all values should be calculated according to what the controller uses internally, typically percentage units):

(1) Set the controller in manual mode. Await (an approximate) steady state and record the control signal (u_{start}) and the measurement (y_{start}). Make a change of the control signal and calculate $\Delta u = u_{final} - u_{start}$

(2) Register the step response of the measurement *y*. Here it is important to wait sufficiently long so that a good estimate of the final value *of* y_{final}

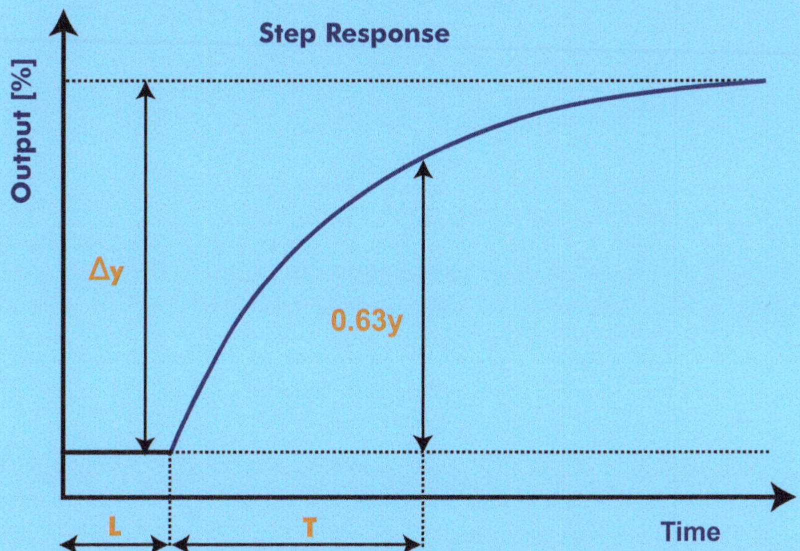

Figure 4.10 Parameters needed from the step response to tune a PI-controller with the lambda method. The time constant is defined as the time it takes to reach 63% of the final value in this first-order system. (Notice that the time to reach 90% of the final value is often used to define the time response of a sensor.)

Table 4.2 Choice of p for the lambda tuning.

p	Controller performance
>3	Slow and stable control
1–3	Relatively fast control
<1	Aggressive control (should normally be avoided)

Table 4.3 Calculation of control parameters.

K	T_i
$\frac{T}{K_S(\lambda + L)}$	T

is obtained. Calculate the increase of the process value and the process gain:

$$\Delta y = y_{final} - y_{start} \tag{4.3}$$

$$K_S = \frac{\Delta y}{\Delta u} \tag{4.4}$$

From the step response (see Figure 4.10) find the time delay L and the time constant T

(3) Choose λ according to

$$\lambda = p * \mathrm{T} \tag{4.5}$$

where Table 4.2 gives guidelines for the choice of p.

(4) Given values of T and K_S (from step 2) and λ (from step 3), the control parameters are calculated according to Table 4.3

Remarks:

- If the process is slow and at the same time is affected by significant disturbances, it may be hard to estimate the required parameters from the step response and we may then have to stick to manual tuning.

An example of lambda tuning in an aeration system is found in Chapter 5.

4.3.4.2 Autotuning

Today most computer control systems are equipped with a feature where the controller is tuned automatically on demand by the user. Such a mechanism is usually called *autotuning* (Åström & Hägglund, 1995, 2006) and typically consists of the following three steps:

- disturb the process;
- evaluate the process response of the disturbance;
- calculate controller parameters.

This is the same procedure that an experienced engineer uses when tuning a controller manually. The process must be disturbed in some way to be able to get information of the process dynamics. This can be done in many ways, for example, by adding steps, pulses, or sinusoids to the process input. A common approach is to use a relay feedback experiment. During the tuning, the controller is then (automatically) replaced by an *on–off* controller that will cause the process output to oscillate. By measuring the amplitude and frequency of the oscillations, the control parameters are calculated using a simple design method. As for all tuning rules, the user needs to give some specifications of the desired control performance. Several autotuning methods have been developed over the last three decades and have proven to be most useful and accepted in the process industry. Autotuning certainly has the potential to be more used in WRRF applications, including aeration control.

4.3.4.3 Integral windup

For a PI (D) controller to work well it is of utmost importance to take into account the fact that all actuators have a limited 'operation interval'. Outside this interval the actuator is saturated, see Figure 4.11. A simple example is that a 'blower' has a minimum and a maximum capacity so that the amount of air that could be fed to the aeration basin is limited. A valve can only be open 100% at most, even if the controller wishes to have a larger flow rate.

Actuator saturation typically occurs during large disturbances, when the controller calculates a larger (or lower) control action than the actuator can deliver. When this takes place the feedback loop is broken and the system runs in open loop because the actuator will remain at its limit independently of the process output as long as the actuator remains saturated. When actuator saturation occurs, the control error is normally not zero. Since the controller cannot eliminate the error as long as the actuator is saturated, the integral term in the PI(D) controller will increase with time and may become large, in other words it

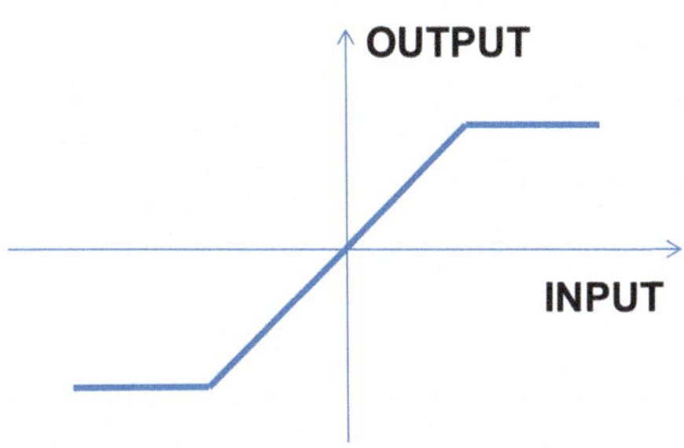

Figure 4.11 A saturator with hard limiting saturation.

'winds up'. When the saturation disappears, the controller cannot work as normal until the control error has been negative for a sufficiently long time. This means that the correction to zero error takes unnecessarily long time. The problem of integral windup is illustrated in Figure 4.12.

The control signal may be limited not only because of actuator saturation. Other reasons include:

- Reduce the consequence of a malfunctioning sensor, that is, independently of the sensor reading, the control signal should stay within a given interval.
- Avoid excessively small or large control signals since they may influence the process negatively. For example, an excessively low aeration rate may make the sludge settle in the aeration basin or increase the risk for diffuser fouling. Furthermore, an excessively low DO may favour the release of nitrous oxide or deteriorate the settling properties of the sludge.
- Often large control signals should be avoided since they are expensive to generate and offer negligible process improvements. For example, increasing the DO level above say 3 mg/l gives only little increase in

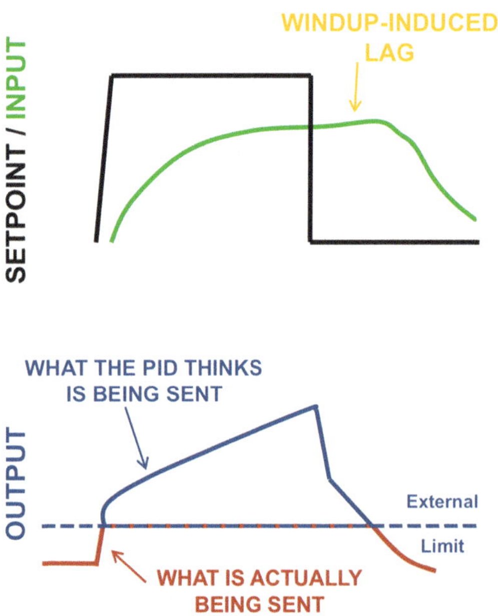

Figure 4.12 Illustration of integral windup. At the start of the load change the controller will see a large error *e*. The PI controller calculates the desired control signal (lower figure) but owing to the actuator saturation the actual control signal is smaller. Therefore, it takes a longer time to compensate for the load change. When the desired control signal is no longer saturated, the error *e* is still positive, owing to the windup. Consequently, the windup causes the correction to be delayed.

treatment efficiency, but costs energy (due to the needed airflow rate). It is not always necessary to correct the DO rapidly. The microorganism activity will not be negatively affected if the DO correction is delayed a few minutes.

Today most control system vendors have systems that can handle integral windup. Below, however, we will just indicate for the interested reader how such handling can be realized.

4.3.4.4 Prevention of integral windup

To avoid integral windup, the integrator part of the controller should be prevented from increasing/decreasing when the control signal hits its upper/lower limit. Today all commercial controllers are provided with anti-windup design (AWD) and the topic has been discussed in the academic literature and textbooks for decades. However, it is still not uncommon to find implementations where integral windup occurs, in particular for cascade control systems.

Basically, for a single controller the user just must select the desired minimum and maximum values of the control signal and feed this information to the control unit where the windup is prevented by the internal AWD. It is absolutely forbidden to limit the control signal by adding an extra function after the controller where the control signal is limited. This will surely lead to integral windup.

The basic idea of AWD is simply to limit the integral part of the controller. Here we just present a simple example of AWD for a single PI controller (controller parameters: K = gain, T_i = integral time, h = sampling interval. The control signal is calculated at times t, $t+h$, $t+2h$, etc.)

$c_1 = K^*h/T_i$ (initial calculation to speed up computation time)
$e = y_{ref} - y$ (calculation of the control error)
$I_{part} = I_{part} + c_1{}^*e$ (the integral is calculated as a sum $K(e(t) + e(t-h) + e(t-2h))$ ...... and I_{part} is set to zero when the control is initiated)
$v = K^*e + I_{part}$ (calculation of the desired control signal)
$u = u_{lim}\ (v, u_{min}, u_{max})$ (the function u_{lim} limits the control signal v between the maximum value u_{max} and the minimum value u_{min})
$I_{part} = u - K^*e$ (anti-windup correction of the integral part. If the v value stays within the limits then the integral part will not be changed by this statement.)

In this way, the integral part is always limited so that the actuator signal equals the desired signal.

4.3.4.5 Prevention of integral windup in cascade control

The idea in cascade control is to have two controllers in series where the output from the first controller, the primary (master) controller, becomes the setpoint for the secondary (slave) controller, see Sec. 4.3. The problem now is that when the secondary control loop goes into saturation, in a standard scheme there is no way for the primary controller to know about this, so integral windup will take place. The solution to this problem is to provide some means of communication between the secondary and primary controllers that allows the secondary controller to inform the primary controller when the output has saturated. Basically, the primary controller should activate its AWD if:

- its control output reaches its limits (this is the same as for the single-loop case)
  or
- the control signal from the secondary controller saturates.

Some logic is needed to get a satisfactory AWD for a cascade control system. If a cascade control is built from two standard PI(D) controllers it is important to make sure that information when the secondary controller saturates is used for the AWD is the primary controller. This kind of feature is usually built in to modern commercial control systems.

4.3.4.6 Example of cascade wind-up

An example of cascade wind-up in the DO controller is shown in Fig. 4.13. The airflow controller loses its control authority when the airflow capacity is not sufficient. The valve position is at its maximum limit during this period, hence the airflow rate and subsequently also the DO concentration cannot be controlled.

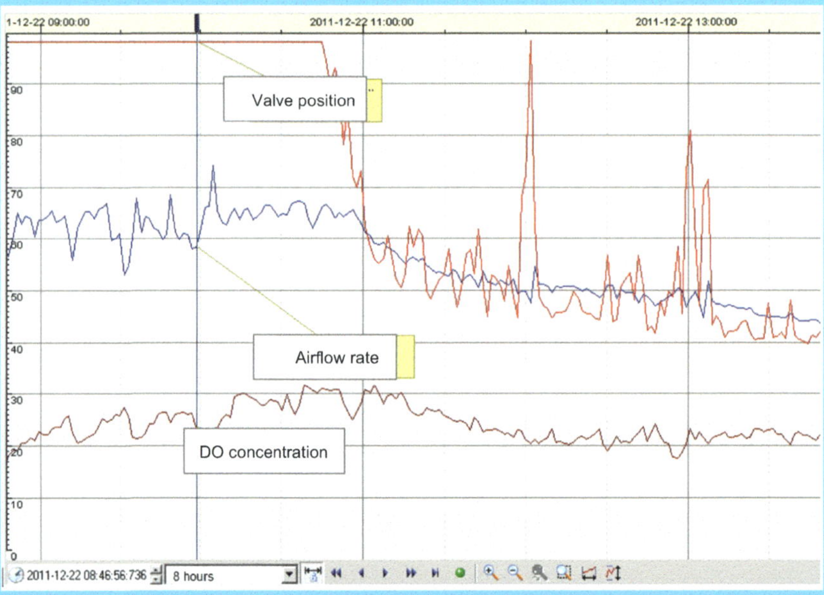

Figure 4.13 Recording of cascade windup of a DO controller.

At the vertical marking in Fig. 4.13, the DO setpoint is reached after a long period of lost control action due to limitations in aeration capacity during a period of high load. It takes an extra hour for the airflow rate to start to go down while the DO concentration crosses 3 mg/l.

Another phenomenon that can be seen on the graphs is that of non-linear valve characteristics. Between 100% and 75% valve opening the airflow rate is not reduced. After this episode, the maximum valve opening was reduced and the airflow and DO controllers were tuned.

4.4 SUMMARY

Even if the need for ICA is no longer called into question, ICA is still perceived as the *hidden technology:* it is noticed when it does *not* work. Certainly, the need for ICA in water and wastewater systems is clear. An ARC Advisory Group study (ARC, 2017) found a fast-growing market for automation and field devices in wastewater treatment applications. ARC believes that the water and wastewater industries represent one of the greatest opportunities for the automation business through the next 20 years. Control theory offers a huge toolbox of methods, sufficient to handle all possible operational conditions in a wastewater treatment system.

4.5 REFERENCES

Åmand L., Olsson G. and Carlsson B. (2013). Aeration control – a review. *Water Science and Technology*, **67**, 2374–2398.

Andrews J. F., Briggs R. and Jenkins S. H. (eds) (1974). *Instrumentation, Control and Automation for Wastewater Treatment Systems*. Pergamon Press, Oxford. (Also *Progress in Water Technology*, 6(5–6)).

ARC (2017). *Water and Wastewater Industry Automation and Software Expenditures*. www.arcweb.com/market-studies/water-and-wastewater-industry-automation-and-software-expenditures (accessed 9 January 2018).

Åström K. J. and Hägglund T. (1995). *PID Controllers: Theory, Design, and Tuning*. Instrument Society of America, Research Triangle Park, North Carolina, USA.

Åström K. J. and Hägglund T. (2006). *Advanced PID Control*. Instrument Society of America, Research Triangle Park, North Carolina, USA.

Åström K. J. and Murray R. M. (2016). *Feedback Systems: An Introduction for Scientists and Engineers*, 2nd edn. Princeton Press, New Jersey, USA. Freely available at: http://www.cds.caltech.edu/~murray/amwiki/index.php/-Second_Edition (accessed 15 February 2018).

Brandt M., Middleton R., Wheale G. and Schulting F. (2011). Energy efficiency in the water industry, a Global Research Project. *Water Practice and Technology*, **6**(2), wpt2011028. doi: 10.2166/wpt.2011.028

Briggs R. (1973). Instrumentation and control in sewage treatment. In Andrews *et al.* (1974).

Briggs R., Knowles G. and Scragg L. J. (1954). A continuous recorder to dissolved oxygen in water. *Analyst*, **79**, 744.

Briggs R., Davies F. S., Dyke G. V. and Knowles G. (1957). Use of wide-bore dropping-mercury electrode and zinc reference electrode for continuous polarography. *Chemical Industry*, **8**, 223–224.

Briggs R., Jones K. and Oaten A. B. (1967). Monitoring and automatic control of dissolved oxygen level in activated sludge plants. In: *Effluent and Water Treatment Convention*. Thunderbird Enterprises, London, UK.

Brouzes P. (1969). Automated activated sludge plants with respiratory metabolism control. In: *Advanced Water Pollution*, S. H. Jenkins (ed.), Pergamon Press, Oxford, UK, pp. 385–396.

Buhr H. O., Andrews J. F. and Keinath T. M. (eds) (1974). Research needs for automation of wastewater treatment systems. In: *Proceedings of a US Environmental Protection Agency Workshop*, 23–25 September 1974. Clemson University, South Carolina, USA.

Collier J. R. (1939). Activated sludge sewage treatment. *Ohio Conf. Sewage Treatment*, **13**, 124–126.

Desborough L. and Miller R. (2002). Increasing customer value of industrial control performance monitoring – Honeywell's experience. In: *Sixth International Conference on Chemical Process Control.* AIChE Symposium Series Number 326 (vol. **98**), pp. 172–191.

Dobbins W. E. (1974). Design of activated sludge plants for computer control. In Andrews *et al.* (1974).

Gillblad T. and Olsson G. (1977). Computer control of a medium sized activated sludge plant. *Progress in Water and Technology*, **9**(5–6), 427–433.

Guarino C. F. and Radziul J. V. (1977). Water–wastewater, instrumentation and automation (I & A), USA. *Progress in Water and Technology*, **9**(5–6), 35–39.

Halvarsson B. (2010). Interaction Analysis in Multivariable Control Systems: Applications to Bioreactors for Nitrogen Removal. *PhD thesis*, Uppsala University, Sweden. Available at: http://www.diva-portal.org/smash/record.jsf?pid=diva2%3A309623&dswid=-622#sthash.mqshlexz.dpbs (accessed 9 January 2018).

Haseltine T. R. (1937). Operating control tests for the activated sludge process. *Waterworks and Sewerage*, **84**, 121–123.

Heukelekian H. (1934). Some biochemical indices of the condition of activated sludge. *New Jersey Sewage Works Association Proceedings*, **19**, 50–55.

Hug T., Benedetti L., Hall E. R., Johnson B. R., Morgenroth E. F., Nopens I., Rieger L., Shaw A. R. and Vanrolleghem P. A. (2009). Mathematical models in teaching and training: mismatch between education and requirements for jobs. *Water Science and Technology*, **59**, 745–753.

Ingildsen P. (2002). Realising full-scale control in wastewater treatment systems using *in situ* nutrient sensors. *PhD Thesis*, Lund Univ., Sweden. Available at: www.iea.lth.se/publications/Theses/LTH-IEA-1030.pdf (accessed 9 January 2018).

Jones K., Briggs R., Carr J. G. and Potter A. H. (1969). Automatic control of aeration in a fully nitrifying activated-sludge plant. *Institution of Public Health Engineers Journal*, **68**, 271–295.

Kampschreur M. J., Temmink H., Kleerebezem R., Jetten M. S. M. and van Loosdrecht M. C. M. (2009). Nitrous oxide emission during wastewater treatment. *Water Research*, **43**, 4093–4103.

Lang L., Pocquet M., Ni B. J., Yuan Z. and Spérandio M. (2016). Comparison of different 2-pathway models for describing the combined effect of DO and nitrite on the nitrous oxide production by ammonia-oxidizing bacteria. *Water Science and Technology*, **75**, 491–500.

Lindblom E., Arnell M., Flores-Alsina X., Stenström F., Gustavsson D. J. I., Yang J. and Jeppsson U. (2016). Dynamic modelling of nitrous oxide emissions from three Swedish sludge liquor treatment systems. *Water Science and Technology*, **73**, 798–806.

Olsson G. (2012). ICA and me – a subjective review. *Water Research*, **46**, 1585–1624.

Olsson G. and Andrews J. F. (1978). The dissolved oxygen profile – a valuable tool for the control of the activated sludge process. *Water Research*, **12**, 985–1004.

Olsson G. and Hansson O. (1976). Stochastic modeling and computer control of a full-scale wastewater treatment plant. In: *Proceedings of Symposium on Systems and Models in Air and Water Pollution*. Institute of Measurement and Control, London, UK.

Olsson G. and Newell B. (1999). *Wastewater Treatment Systems. Modelling, Diagnosis and Control*. IWA Publishing, London, UK.

Olsson G., Carlsson B., Comas J., Copp J., Gernaey K. V., Ingildsen P., Jeppsson U., Kim C., Rieger L., Rodríguez-Roda I., Steyer J.-P., Takács I., Vanrolleghem P. A., Vargas Casillas A., Yuan Z. and Åmand L. (2014). Instrumentation, Control and Automation in wastewater – from London 1973 to Narbonne 2013. *Water Science and Technology*, **69**, 1373–1385.

Palmer J. R. (1939). Activated sludge control. *Ohio Conference on Sewage Treatment*, **13**, 21.

Pearse L. (ed.) (1941). *The operation and control of activated sludge sewage treatment works*. Report of Committee on Sewage Disposal. American Public Health Association (APHA), Atlantic City meeting of the APHA, Oct. 14, 1941. *Sewage Works Journal*, 14, 3–69.

Petersack J. R. and Stepner D. E. (1973). Computerized data management and control of a secondary wastewater treatment plant. In Andrews *et al.* (1974).

PID Loop Tuning Pocket Guide (2017). Control Soft, Version 5, available at: http://www.controlsoftinc.com/pocketguide.shtml (accessed 9 January 2018).

Rieger L., Alex J., Gujer W. and Siegrist H. (2006). Modelling of aeration systems at wastewater treatment plants. *Water Science and Technology*, **53**(4–5), 439–447.

Roesler J. F. (1974). Plant performance using dissolved oxygen control. *Journal of Environmental Engineering*, **100**(EE5), 1069–1076.

Rosso D., Stenstrom M. K. and Larson L. E. (2008). Aeration of large-scale municipal wastewater treatment plants: state of the art. *Water Science and Technology*, **57**, 973–978.

Torpey W. N. (1948). Practical results of step aeration. *Sewage Works Journal*, **20**, 781.

Townsend-Small A., Pataki D. E., Tseng L. Y., Tsai C. Y. and Rosso D. (2011). Nitrous oxide emissions from wastewater treatment and water reclamation plants in southern California. *Journal of Environmental Quality*, **40**, 1542–1550.

WEF 2009. *Energy Conservation in Water and Wastewater Treatment Facilities*, WEF Manual of Practice [WWW Document], McGraw-Hill, Inc., New York, NY, USA. Available at: http://www.amazon.com/Energy-Conservation-Water-Wastewater-Facilities/dp/0071667946 (accessed 25 February 2015).

Ziegler J. G. and Nichols N. B. (1942). Optimum settings for automatic controllers. *Transactions of the ASME*, **64**, 759–768.

Chapter 5

Aeration control – implementation

Gustaf Olsson[1], *Linda Åmand*[2], *Leiv Rieger*[3] *and Bengt Carlsson*[4]

[1]*Lund University*
[2]*IVL Swedish Environmental Research Institute*
[3]*inCTRL Solutions Inc.*
[4]*Uppsala University*

5.1 INTRODUCTION

After presenting the fundamentals of control for aeration and dissolved oxygen control, this chapter focuses on the implementation issues. The early years of DO control are described in Chapter 4. Since the 1970s a big effort has been directed towards improving DO control, driven by the desire to reduce the costs induced by this energivorous process (Olsson, 2012). The state-of-the-art to 2005 was summarised in Olsson *et al.* (2005). A review of aeration control with emphasis on the 21st century is found in Åmand *et al.* (2013). Successful DO control depends on many more factors than algorithms. The key components of the control systems are described in Sec. 5.2. The challenges facing the implementer are discussed in Sec. 5.3, from defining the goal of the control to the various stages of a successful realization of the control, before Sec. 5.4 presents a brief outlook.

Editor: Diego Rosso
doi: 10.2166/9781780407845_0143

5.2 COMPONENTS OF A CONTROL SYSTEM

An aeration control system relies on many components. There is a whole chain of equipment that must work together for a DO control system to be functional. Pressurized air is produced by a blower/compressor. A pressure control system must guarantee that the pressure is adequate under the dynamic changes of the load. Air valves must be controllable so that the air distribution along the aerator can satisfy the required airflow. Nutrient and DO sensors must provide information from the correct locations. Airflow measurements should inform the airflow controller how much to change the airflow. Valve positions ought to be measured so that the most open valve all the time can be as open as possible, and that the pressure drop over the valves is minimized. In other words: DO control includes much more than DO sensors and valves.

5.2.1 Sensors in aeration control

An important development in nutrient sensors has taken place in the last two decades, from automated laboratory *ex situ* analyzers that had to be protected from the measured system through filtration units to *in situ* sensors that can be placed directly in the liquid to be monitored. Advances include optical sensors based on luminescence techniques for DO measurements requiring less maintenance compared with membrane-based sensors. Online *in-situ* nutrient sensors are becoming common and affordable, for example, ion-selective electrode (ISE) probes for ammonia (Rieger *et al.*, 2002) and UV probes for nitrate and nitrite (Rieger *et al.*, 2004). This has eliminated long measurement delays and slow sensor dynamics, resulting in easier control and better performance. However, a recent trend has been a renaissance in classic *ex-situ* analyzers caused by improved filtration units and modernized analyzers with significantly reduced usage of consumables. To find out if a sensor is 'affordable' we must consider both maintenance cost and payback time. Sensor maintenance is a key factor and is getting increasingly favourable, so the payback time for sensors is decreasing for every year.

The idea of this section is not to provide technical details of all available sensors but rather to focus on availability and practical applicability of sensors for aeration control. Table 5.1 presents an overview of key sensors for aeration and DO control. For details on the measuring principles the interested reader is referred to standard textbooks like WEF's *Manual of Practice 21* (WEF, 2013).

Quality Control is an often overlooked part of operating online sensors but is required to guarantee proper functioning of any control system. It is recommended to develop a detailed quality control and maintenance plan including training of instrumentation staff and a clear incentive system based on easy-to-understand metrics.

Table 5.1 Overview of key concentration sensors with respect to their availability, applicability, maintenance and requirements.

Dissolved oxygen (DO)	There are two main principles for DO sensors: membrane and optical probes. The trend is towards optical probes due to the reduced maintenance requirements. However, membrane-type probes may have advantages in environments with low DO levels. The main maintenance activity will be to clean the probe. Automatic cleaning devices are recommended. For membrane probes the membrane must be exchanged and the electrolyte must be refilled on a regular basis. Optical probes typically need a replacement cap once a year. Re-calibration is typically only required few times a year. However, the accuracy should be regularly checked as part of the quality control program. A comparison with a well maintained and calibrated DO handheld device is recommended at least once a week. Special attention should be given to designing quality control procedures which cover the whole measuring range and not simply generate a cloud of control points around the DO setpoint. *Special note*: Most DO probes use a calibration curve artificially changed by the manufacturer, which introduces a negative zero DO point. Negative values are then cut off. This should make sure that the probes are showing zero DO in unaerated reactors and not concentrations of, say, 0.1 or 0.2 mg DO/L. While this has limited impact in normal operation, it can introduce control problems when low DO setpoints are targeted (e.g., by an ABA controller or in short-cut nitrogen removal systems). This also causes an additional burden for quality control as the zero point cannot be easily checked.
Total ammonia NH_x	Total ammonia is the key variable to measure the nitrification performance. It can be expressed as NH_x-N (sum of NH_3-N and NH_4-N). Most ammonia at WRRFs is in the form of NH_4-N with only a small portion as NH_3-N. However, remember that the pKa of ammonia is 9.2, hence at high pH (about >8) and temperature the ion equilibrium changes to a significant level of the unionized form (NH_3-N). The following methods most applied: (1) colorimetric, (2) gas sensitive electrodes (measuring NH_3-N after raising the pH with the result that all ammonia is in the form of NH_3), and (3) ion-selective electrodes (ISE, measuring NH_4-N). NH_x probes using the ISE approach need to calculate total ammonia based on additional pH and temperature sensors if pH is higher than ≈8 (depends on targeted accuracy and temperature/pH combination). Methods (1) and (2) typically require a filtration step before the actual measurement.

(*Continued*)

Table 5.1 Overview of key concentration sensors with respect to their availability, applicability, maintenance and requirements. (*Continued*)

	In situ ISE probes can be applied to primary effluent or installed in a biological reactor. Low total ammonia concentrations (i.e., at the end of the aerated zone) may be difficult to measure with some ISE probes and manufacturer specifications should be followed (Cecconi *et al.*, 2018). The response time (ISO 15839, 2003) of NH_x *ex situ* analyzers is typically around 15 minutes, which is sufficient for aeration control. *Ex situ* analyzers require regular maintenance consisting of thorough cleaning of the filtration unit, and replacement of moving parts (e.g., internal tubes) and chemicals for the analyzer. Modern analyzers will have reduced the required maintenance to a minimum: however, it is still a major demand on time and effort and this labor intensity may be a deciding factor against such equipment. *In situ* ISE probes require the regular replacement of membranes and electrolyte, both for the reference and the measuring electrode. Some probes also measure interfering ions and these electrodes need to be maintained as well. ISEs drift over time and need regular re-calibration (sometimes referred to as matrix adaptation). The strongest interfering ion is potassium, which may have a significant influence on the accuracy of the instrument (Cecconi *et al.*, 2018). Ammonia analyzers and especially ISE probes need regular (at least once per week) quality control, as their proper functioning is critical to the performance of any ammonia-based aeration control or nitrification performance monitoring. Special attention should be given to designing quality control procedures that cover the whole measuring range. This is often an issue in fully nitrifying systems when measuring at the end of the aerated reactor, as the concentration will be low most of the day and sometimes close to or below the limit of detection. In these situations, special experiments using sample spiking (adding a highly concentrated standard and measuring the resulting concentration in the laboratory) or proper standard addition experiments (e.g., Funk *et al.*, 2007) should become part of a standard quality control procedure.

	The selection of the total ammonia instrument is not an easy question to answer. *Ex situ* analyzers typically provide a higher accuracy and have auto-calibration routines. On the other hand, ISE probes need less maintenance and often provide sufficient accuracy for ammonia-based aeration control. In general, there is a shift of effort from maintenance (*ex situ* analyzers) to quality control (*in situ* sensors). Critical are situations where the accuracy of the ammonia sensor is not sufficient for the application. Some installations have mitigated the insufficient accuracy of an ISE ammonia probe by moving the probe further upstream to higher concentration ranges. However, this limits the efficiency of the controller, as safety factors for the uncontrolled volume after the NH_x probe must be applied.
Nitrates NO_x	Most sensors measure the sum of nitrite (NO_2-N) and nitrate (NO_3-N), often expressed as NO_x-N. Most nitrogen will be in the form of nitrate (NO_3-N), however, nitrite accumulation may occur under special loading situations or conditions or in specific processes. NO_x can be measured by: 1) optical sensors or analyzers and 2) colorimetric (*ex-situ* analyzer with filtration required). Optical sensors or analyzers measure the absorbance of UV light and correlate it to an NO_x-N concentration. *In-situ* sensors require a method to compensate for any turbidity and organic matter interferences. Optical analyzers typically use an upfront filtration and therefore do not need to compensate for particles. Colorimetric analyzers are rarely seen in wastewater applications. *In-situ* optical NO_x sensors can be installed at a variety of locations in a WRRF. According to the targeted measuring range, different instrument configurations and calibration models may apply. All optical instruments must rely on clean measuring windows, so auto-cleaning devices are highly recommended. Special care should be given to chemical precipitation on the windows (e.g., iron). These deposits may not be visible to the bare eye and may require cleaning using a special base or acid cleaning solutions.

(*Continued*)

Table 5.1 Overview of key concentration sensors with respect to their availability, applicability, maintenance and requirements. (*Continued*)

	The sensor measurements should be compared to the sum of nitrite (NO_2-N and NO_3-N measured with a reference method in a laboratory. Comparing to only nitrate may be misleading. As for all optical probes, special attention should be given to the frequency of manual and automatic cleaning. This can easily be done by comparing measurements before and after cleaning. It is recommended to plot the difference over time to get a better understanding of the drift caused by sensor fouling.
Nitrite NO_2	Probes are available for the measurement of nitrite and are getting increased interest for new processes like anammox, or short-cut nitrogen removal processes.
Nitrate NO_3	Nitrate and nitrite can also be measured with an ion-sensitive electrode (ISE) with chloride being the main interference for nitrate. Nitrate ISE probes may make sense in cases where a multi-electrode probe is applied (e.g., NH_x and NO_3).
ORP	ORP (or redox) stands for oxidation–reduction potential, which is a measure of the tendency of a chemical substance to oxidize or reduce another chemical substance. An ORP sensor consists of an ORP electrode and a reference (chemically inert, platinum) electrode, in much the same fashion as a pH measurement. The sensing electrode potential is read relative to the reference electrode and the value is presented in millivolts (mV). When used in wastewater treatment systems, the oxidation–reduction potential is a measurement of the ability or potential of wastewater to permit the occurrence of specific biological (oxidation–reduction) reactions. Such reactions in wastewater treatment systems include nitrification, denitrification, biological phosphorus removal, and BOD removal. The ORP often displays a 'knee' or a breakpoint in anoxic operation.

	The ORP is an indicator of the sum of many phenomena and is not a universal unique number for all plants. Instead it must be established empirically for each plant. In that way, it can give significant information.
TSS	Total suspended solids refer to the content of solids that are not fast settling out of the water. Insoluble particles of soil, organics, microorganisms and other materials impede the passage of light through water by scattering and absorbing the rays. This interference of light passage through water is referred to as turbidity. Commercial turbidimeters (i.e., nephelometers) for determining low turbidity measure the intensity of light scattered at right angles to the incident light. Nephelometric Turbidity Units (NTU) are sometimes referred to as FTU, since Formazin polymer is used as the reference turbidity standard suspension. There is an international standard for turbidity (ISO 7027). At high turbidity levels, the instrument combines signals from the forward scattered and transmitted light detectors in a denominator of the ratio, allowing linearity over the entire range. Many sensors for MLSS concentration in the aeration basin and the return sludge are based on the near-infrared (NIR) measuring technique. Typical ranges for raw wastewater are 0–1000 mg/l and 0–200 mg/l for primary and secondary effluents. Air bubbles and temperature are known to interfere with the measurement; some sensors have different ways of compensating for this interference.

5.2.1.1 Airflow measurement

A DO control system also depends on the measurement of physical variables, primarily the airflow rate. The basic principle of airflow measurement is simple: the cross-sectional area of the pipe is multiplied by the average velocity of the air. This gives the flow rate. The difficult part is to precisely measure the air velocity across the pipe. The air is slowed by friction with contact to the surface of the pipe. The greatest velocity under laminar flow conditions is in the middle of the cross-section.

The simplest airflow measurement principle is based on differential pressure, the Venturi meter. By measuring the differential pressure across a pipe restriction, the airflow can be estimated. Such a flowmeter uses Bernoulli's equation to measure the flow of fluid in a pipe. A constriction in the pipe creates a pressure drop across the flowmeter. Bernoulli's equation states that the pressure drop across the constriction is proportional to the square of the flow rate. When the flow increases, more pressure drop is created. The differential pressure between the upstream and downstream air is measured and will determine the airflow. This technology accounts for one-fifth of the world market for flowmeters.

The upside of this technology is low cost. It is a well understood way to measure flow. On the negative side is that the measurement range is not good due to the non-linear relationship between the differential pressure and flow rate.

5.2.1.2 Installation, configuration, initialization, and base calibration

The installation of a sensor is probably the most important step towards reliable and robust measurements. It is therefore recommended to follow a detailed initialization procedure. This procedure makes sure that accurate measurements are collected before the sensor is used in any application. This is time well spent as it guarantees proper installation and calibration from the beginning and saves later painful searches for sensor misreadings. The following steps are suggested:

- *Installation* according to manufacturer specifications.
- *Sensor configuration* and *connection to SCADA/DCS system* including setting parameters according to the specific installation.
- *Test of sensor integrity*: Analysis and optimization of signal-to-noise ratio, review of installation, cleaning procedures and intervals, etc. Often changing the sensor location or the specific installation (e.g., changing the angle of how an *in situ* sensor enters the water) may make a significant difference. Supplier advice should be complemented by trial and error for installations with sub-optimal results.
- *Collection of a dedicated data set* in the form of grab samples analyzed in a laboratory and compared with the sensor measurements. The data set should

cover the targeted measuring range and will be used as a basis for a detailed analysis of the sensors accuracy (compare Rieger *et al.*, 2005). Special experiments may have to be designed to collect this data set.

- *Test of application integrity*: Check of calibration and response time on sensor and SCADA/DCS level. Note that for an application in a control loop, the SCADA value is used and therefore the signal transmission needs to be included in an evaluation. Acceptance of existing default calibration or local base calibration (e.g., matrix adaptation or simple 2-point calibration)
- *Advanced base calibration* (e.g., full calibration or manufacturer service), if required
- *Base calibration* is the process of deriving a calibration curve, which converts measured raw signals (e.g., mV signal) into a concentration, flow or any other variable of interest. A Base Calibration should be based on a dedicated calibration data set resulting from specifically designed experiments, where highest accuracy standards have been applied.

5.2.1.3 Maintenance

Maintenance of sensors and other control equipment is expensive and takes away from potential savings achieved by an advanced control solution. It is recommended to develop a detailed maintenance plan for all sensors, which coordinates all activities and aims at finding a balance between data quality and maintenance efforts and costs.

When setting up a maintenance plan for a sensor, the first step is to follow the manufacturer recommendations and in special applications consult with a supplier's technician. In a second step, the experience of operators and maintenance personnel should be collected to complement the manufacturer recommendations. Sensor maintenance can be categorized into: (1) cleaning, (2) replacement of parts or reagents, and (3) maintenance calibration.

- *Cleaning*: Fouling or clogging of sensors accounts for most sensor faults. Making sure that the instrument is cleaned regularly and according to the location and installation conditions is therefore the key to accurate sensor readings. The specific environment in which the instrument is installed determines the frequency and methods of cleaning. Cleaning may simply consist of the wiping of an *in situ* sensor or the time-consuming chemical cleaning of a filtration unit or of the fine tubes in an analyzer.
- *Replacement of parts or reagents*: Most instruments require exchange of parts or refilling of chemicals or reagents on a regular basis. The replacement intervals are typically provided by the manufacturers but should be complemented by the plant's own experience with the instrument at its specific location. Some parts are best replaced by the manufacturer, while others can be easily replaced by operators.

- *Maintenance calibration*: This is the process of resetting the sensor to measure according to the specifications. Often the aging of parts or chemicals is compensated (e.g., membranes, electrolyte, reagents, etc.). It should not be confused with a base calibration, which is the process of establishing a calibration curve to relate the measured raw value (e.g., a mV signal) to concentration measurements as described above. Excessive maintenance calibration indicates that the sensor needs a detailed review by an instrumentation expert. Adjustments to an established base calibration need to be based on a proper database. The overly zealous forcing of a sensor to match what is perceived to be correct based on observed process performance would defeat the purpose of sensors as a reliable source for measuring process performance.

5.2.1.4 Quality control

Quality Control is the act of comparing online sensor readings during normal operation with other sensors (e.g., handheld devices) or reference measurements typically based on grab samples analysed in a laboratory using standard methods. Quality Control should guide detailed inspections and, if required, recalibration.

As a rule of thumb, reference measurements should be taken once a week or more often depending on the risk of a faulty measurement. Additional quality checks may be applied to every measurement entering a control system using advanced methods from statistical quality control (e.g., Montgomery, 2009).

For the analytical instruments described above, it is recommended to involve trained laboratory personnel in the quality control and calibration work. However, it is good practice to develop tools and standard operating procedures to enable operations personnel to carry out quality control of the sensors under their responsibility. These tools should provide graphical feedback of the instrument's status for instance in form of control charts.

5.2.2 The air supply system: actuators

The actuators - valves, motors, pumps and compressors – are crucial parts of a control system. Inadequate actuators may limit the ability for control, that is, by having a too low maximum or too high minimum capacity. This is a fundamental barrier for more widespread acceptance of new control strategies. Many existing wastewater treatment plants are simply not designed for real-time control (Ingildsen, 2002).

5.2.2.1 Blower air supply

Variable-speed control of motors can provide smooth pumping and easily adjustable compressors. In the 1970s and 1980s variable speed control was a luxury. For air compressors, guide vane control could only provide limited control authority, resulting in less accurate DO control. Olsson (1985) pointed out that valves were (and still are) often not designed for good control.

Today variable speed control is a proven technology because of the development in power electronics over the last few decades, from mW scale motors to MW drives. Power electronic devices like IGBT (insulated-gate bipolar transistors) are now generally available for currents up to 1200 A and voltages up to 3000 V. Variable speed control has a profound influence on wastewater treatment operations in flow rate control as well as for airflow control (for details on blowers and their operation and applicability, see Chapter 1).

The pressure control of the air supply via various control actions on the blower is coupled to the DO control. If the DO control system increases the airflow to the aeration basin, then the pressure in the air supply system will drop. This is compensated by the pressure control system of the blower that will change the air production from the blower. Similarly, if the DO control system decreases the airflow then the pressure will rise and the pressure controller will change the blower air production. The two control systems are not directly coupled via any electronic system: they are physically coupled via the air pipes.

5.2.2.2 Air distribution control

In a plant with parallel aeration basins the air requirement is seldom the same in all the aeration grids. This means that the air valves to the different grids have different openings. Any valve that is not fully opened will cause a pressure drop, which simply wastes energy but does not affect the airflow. Therefore, it is logical to try to keep the most opened valve almost fully open. To achieve this, it is necessary to measure the air valve opening. This approach enables the control of the DO according to 'the most open valve' control method (Åmand *et al.*, 2013; Olsson & Newell, 1999). DO control will require an increase or decrease of the airflow rates into the aerators. If the most opened valve is not fully open it means that energy is wasted. By decreasing the pressure setpoint (that is no longer constant) all the air valves must open a little more.

Now we see that both the DO control system and the blower pressure control can influence the valve openings. To avoid confusing the control system, the pressure setpoint cannot be changed as quickly as the DO changes. Instead, the airflow rate is changed typically every 5–10 minutes. The pressure control system will compensate for these changes. However, to change the pressure setpoint must be done much more seldom, typically not more frequently than every 2–4 hours. Working in different time scales we will avoid any strange coupling between the control systems.

An additional advantage of monitoring the valve openings is using it for maintenance purposes. By comparing the valve opening to an airflow (or a liquid flow) it is possible to detect a pipe clogging or increased friction in the valve operation.

5.2.2.3 How frequently should you change the airflow?

It is crucial to consider the dynamics when closing the loop. Sometimes controllers are tuned to be too 'ambitious'. For example, a DO sensor may show different DO concentration values every 10 seconds. The reason is that the oxygen uptake rate can change quickly, depending on the mixing behaviour or on quick variations of the load. This does not mean that the airflow rate should be changed so often, since the typical response time of the DO concentration from an airflow change in a full-scale aerator is 15–30 minutes. Frequent changes of the airflow – more often than every minute – will only produce meaningless control actions and wear out the actuators, which would just keep hunting for a position based on this high-frequency (*de facto*) noise (see description in Chapter 1). Instead a control action every 5–10 minutes is more than adequate. DO measurements may be made every 10 seconds and then averaged over several minutes.

5.3 IMPLEMENTING AERATION CONTROL

Here we describe a systematic approach for the design, implementation and evaluation of aeration control solutions, according to Figure 5.1. The concept of

Figure 5.1 Proposed method for implementing and maintaining a new or updated aeration control solution.

control solution includes a control strategy, control algorithms, necessary instrumentation and actuators and the placement of those as well as maintenance plans for the full control solution. The process includes:

- *Definition of control goals and analysis of current status:* this is an iterative process where the control goals are defined and data about status is collected and analysed. Constraints revealed during the data analysis may call for the definition of objectives to be updated.
- *Design and preparation*: here decisions are made concerning what control solution to choose to meet the control goals for a range of scenarios. Preparations include deciding on necessary changes to equipment or software.
- *Implementation* involves installing necessary equipment and programming the new control solution into the control system.
- *Operation and maintenance*: this comes after successful implementation.
- *Performance evaluation*: is done after a period of operation and maintenance.

5.3.1 Definition of control goals and constraints

The general definition of a control goal was discussed in Chapter 4. Here we will consider this specifically for aeration systems.

5.3.1.1 Definition of control goals

The *control strategy objectives* are the overall motivations for implementing a new control strategy. In contrast, *control goals* should be defined as specifically as possible, focused on the individual controller development and with the goal to maximize, minimise or balance the individual processes. In aeration control, control goals are often related to the effluent nitrogen concentrations, energy or other resources, and the performance of aeration equipment.

When designing control solutions, it is good practice to first analyse the impacted processes. A good control solution tries to provide optimal conditions for the processes it targets. For instance, to reduce aeration energy consumption one control solution would be to lower the DO setpoint until the overall nitrification performance is limited so that the average ammonia effluent concentration is just below the effluent permit. Another common control solution is to lower the DO setpoint in a downstream section of an aeration tank. These steady-state designed control solutions will fail when the load situation changes or the temperature drops. A better solution is to define an ammonia controller, which is defining how much air is required in a dynamic way. It will automatically set the nitrification performance required. At high ammonia load the nitrification capacity will be maximized, while in low load situations, the aeration intensity is reduced.

5.3.1.2 Definition of constraints

Before finally deciding the control goals, an inventory of *constraints* should be made. Constraints at the plant may (i) prevent the control strategy from reaching the control goal, or (ii) define an additional boundary that will overwrite control actions. There are different types of constraints that can limit how the control goals can be formulated or what solution to choose to implement. A list with examples of different constraints is found in Table 5.2.

- *Regulatory constraints:* The effluent criteria define what effluent limits must be reached, if the limit should never be exceeded or if an average concentration is compared to the limit.
- *Equipment constraints:* Typical equipment constraints in aeration are blower maximum capacity and turn-down capabilities, airflow requirements for mixing, airflow per diffuser, and valve or piping constraints.
- *Process constraints:* Process constraints limit the treatment performance in the process. A typical example is that at high load situations the mass of specific organisms (e.g., nitrifiers) is not sufficient to convert certain pollutants down to a targeted level. With respect to nitrification this means that the nitrification process is limited by the mass of active nitrifiers and not by the aeration intensity.
- *Organizational constraints:* Not all control strategies can be properly implemented and maintained at any treatment plant. The plant may have constraints related to cost of investing in new equipment, or experience level of the plant personnel, or a clear incentive system is missing.

5.3.1.3 Prioritization of control goals

A challenge when defining the control goals is to set up *priorities* of the respective control goals, taking the constraints into consideration. Prioritizing between control goals can be a tricky task, since goals and/or constraints can be in conflict with each other. As an example, if effluent nitrogen were not a high priority, shutting down aeration would be the most energy-efficient solution.

Regulatory constraints have the highest priority. BOD or nitrogen limits in the effluent permit must be met. Regulatory constraints should be translated to control goals by defining the production goals for the plant, for example, keep on average 2 mg/l below effluent permit, or never be closer than 1 mg/l below effluent permit.

Next, define the control goals other than those related to the regulatory constraints. Common control goals are to reduce energy cost, reduce energy consumption, reduce methanol costs, increase plant treatment capacity, or reduce costs due to manual operation. If the carbon footprint is under discussion at the plant, minimizing nitrous oxide emissions could also be a control goal. Set priorities between the control goals. This is an easier task if a proper data collection and analysis was made.

Table 5.2 Examples of regulatory, process and organizational constraints that impact on how the control goals can be formulated. The examples are focused on plants with nitrogen removal.

Type of Constraints	Example of Constraint	Impacts of Constraints
Regulatory constraints	TN below 10 mg/l (annual) Ammonia below 2 mg/l (monthly) Never to exceed limit on TN	Minimum nitrification and/or denitrification rate(s) on long or short term
Equipment constraints	Blower maximum capacity limit Blower turn-down Maximum wasted sludge mass due to constraint in sludge handling	Maximum nitrification and/or denitrification rate(s)
	Minimum airflow rate through diffusers	Minimum nitrification rate
Process constraints	Maximum solids concentration due to limitations in settler capacity Mass of active nitrifiers (depends on average ammonia removed)	Maximum nitrification rate
	Minimum airflow to keep sludge in suspension	Minimum nitrification rate
Organizational constraints	The budget limits investments in new equipment Time plan Knowledge of process control within the organization/operating team	Choice of control solution to meet the control goals

5.3.1.4 Time plan and budget

In this step of the process, it is also important to consider the time plan and budget. Discuss when the implementation needs to be finalized, important milestones along the way and what the budget is, including budget for investments during implementation and maintenance during operation. A limited budget or time plan may impact the choice of control solution.

5.3.2 Data collection and analysis

5.3.2.1 Data collection

Collection and understanding of data and information is a cornerstone when making informed decisions on the new control solution. Plant-specific data should be collected that might constrain or define the control solution. This includes looking at:

- *influent data* – including temperature variations, flow variations, load variations to the plant and to the bioreactors,
- *effluent data* – including effluent peaks and seasonal variations,
- *process data* – such as limitation in sludge age, sludge bulking problems, nutrient-rich side streams added to the bioreactors, flow distribution to the bioreactors, and
- *equipment data* – such as blower capacity, diffuser density, how present controllers and actuators function.

5.3.2.2 Discuss plant specifics

A key question is whether the plant is operating approximately at its design load or if the plant is over- or under-loaded (an assessment of this could be a measurement of the OUR via off-gas or respirometry, for example). These situations create different challenges with respect to control. A plant which is operating at lower capacity than design load might have problems with over-aeration. The aeration system as well as the bioreactors might be larger than necessary. If the plant performs nitrogen removal, complete nitrification might be reached, even though this is not necessary. *In general, it is easier to operate a process that is over-loaded than a process that is under-loaded.*

From a control point of view, the goal is often to limit aeration to save energy by oxidizing less ammonia. For plants operating at their design load or even above, there might be a smaller margin to the effluent criteria. Due to restrictions in SRT due to settler capacity or sludge treatment limits, the nitrification capacity compared to the load could be an issue. Disturbance rejection is probably more crucial than if the plant were under-loaded.

After data collection and analysis, the control goals need to be revisited and possibly updated.

5.3.3 Design and preparations

It is now time to decide which control solution should be implemented in the control system, and prepare for implementation at the plant. Based on the definition of objectives and data collection and analysis the new control solution can be formulated. In most cases, we suggest a step-wise procedure where

- a first screening is made,
- model-based analysis evaluates different control solutions at different flow, load and temperature situations,
- there is discussion with the client to reach a decision on the solution to be implemented.

A major benefit of using models to design the control solution is that a model offers the possibility of testing many different scenarios without making changes to the real process. This means that testing at the plant can be narrowed down to the most promising control solutions. If using a model for controller design, additional data might be needed for model calibration. A recommended read are the *Guidelines for using activated sludge models* (Rieger *et al.*, 2012) which has recommendations on models for aeration control design.

The design process involves deciding on the control structure, control algorithms and controller tuning, see Table 5.3 for the most commonly used options. Depending on the control goals, the complexity of the control structure and the control algorithm may vary. *Always choose the simplest possible control solution that meets the control goals.* It is the control structure that defines the necessity of new equipment in the process as well as future maintenance needs of the equipment. When the pairing of inputs and outputs is done, similar control behaviour can often be reached with different choices of control algorithms (Stare *et al.*, 2007).

5.3.3.1 Control structure and algorithms

A guide to the most common control structures is found in Table 5.3, providing an overview of possible benefits and drawbacks of the respective control structures and what opportunities the controller settings may offer (excluding tuning of the control algorithm).

In Table 5.3, the least-complex control structure is DO control with or without airflow rate control. Is the airflow rate controller really needed, since airflow sensors cost a lot? As indicated before, an important advantage of a slave controller for airflow rate is that it linearizes the non-linear valve curve thanks to cascade control. It also makes it possible to set limits of the airflow rate which offer safety in failure. Setting airflow limits provides the possibility to take into consideration both the manufacturer's lowest airflow rate per diffuser and the airflow rate needed for mixing. Also, should the DO sensor fail, there is always a limit on the minimum airflow rate directed to the basin. Instead, without slave controllers the valve opening cannot be modified.

Table 5.3 Potential benefits and drawbacks of common control structures in aeration control, and opportunities with the controller settings.

Control Structure	Potential Benefit	Potential Drawback	Settings	Opportunities thanks to Settings
DO control without airflow control	Energy saving, increased denitrification More benefits if several compartments use DO control Reduced installation cost when airflow sensor is left out	Close monitoring of effluent concentrations required Lack of airflow control may make tuning trickier, LL and HL on airflow rate cannot be specified and lack of fallback strategy if DO controller fails.	Valve LL Valve HL	Avoiding the most nonlinear valve characteristics
			DO SP	Estimated DO setpoint required to meet control goal(s)
DO cascade control with airflow control	Energy saving, increased denitrification. More benefits if several compartments use DO control If airflow is controlled, airflow limits can be set.	Close monitoring of effluent concentrations required	Valve LL Valve HL	Avoiding the most nonlinear valve characteristics
			Airflow rate LL	Manufacturer recommendation on minimum airflow rate through diffusers, aeration for mixing
			Airflow rate HL	Maximum airflow capacity through diffusers
			DO SP	Estimated DO setpoint required to meet control goal(s)

Ammonia cascade control	Energy saving, increased denitrification, reduced external carbon dosage, less manual operation	More sensors to maintain, slow reaction to ammonia change if ammonia concentration varies fast	DO LL	Potential nitrous oxide formation at low DO concentrations
			DO HL	Maximum DO that can impact nitrification rates
			Ammonia SP	Safety margin to effluent criteria
Volume control	Nitrification capacity can be increased quickly to reduce ammonia peaks	Requires swing zones, careful tuning to avoid oscillations in effluent ammonia when swing zones are switched on/off	Depends on control algorithm	
FF ammonia control	Faster response to a disturbance	More complex controller, process model required, more sensors to maintain, possible limited impact if there is a kinetic constraint, requires complementary feedback control	Depends on control algorithm	

FF = feedforward, LL = Low limit, HL = High limit, SP = Setpoint.

Similarly, if an ammonia controller controls the airflow rate directly instead of including a cascaded DO controller, a possible fallback strategy is lost and a DO concentration limit cannot be set. A back-up strategy in this case is of the utmost importance as a DO limit is always necessary to avoid unnecessary aeration at ammonia peaks. When the DO concentration in the whole aerated volume is around 2 mg/l, there is no additional impact on the nitrification rate from increasing the DO concentration further. The nitrification rate is at this point not oxygen-limited but limited by the amount of nitrifiers in the system. The tricky part is knowing when the whole aerated volume has a DO concentration of 2 mg/l. The DO sensor might not be placed at a representative location, which is why trial and error on the best maximum DO concentration in the ammonia controller may be necessary.

If the full potential of energy saving with ammonia control is to be exploited, the system should only be limited by the maximum DO concentration in the ammonia controller. On the other hand, limiting the system by the minimum DO offers an opportunity to save energy by setting the lowest allowed DO concentration in the process. Care must be used, however, since the metric for a healthy process is not the DO setpoint but the avoidance of large DO gradients within the reactor volume: as we will discuss later in the book, a process can be functioning well with a lower DO value *throughout* the reactor volume, while a higher DO setpoint guarantees that the DO is above the 'dangerous' levels for all zones within a reactor.

Furthermore, much effort has been made during the last two decades to understand potential nitrous oxide (N_2O) emissions from biological wastewater treatment. N_2O is a strong greenhouse gas with a global warming potential about 300 times that of CO_2 on a mass basis. Low DO concentrations in the nitrification and denitrification processes have been reported as a risk factor that could lead to nitrous oxide emissions (Foley *et al.*, 2010; Kampschreur *et al.*, 2009). Low oxygen levels, sudden changes of ammonia load and higher temperatures might cause nitrite concentrations to increase (Foley *et al.*, 2010). High nitrite concentrations lead to the reduction of nitrite to nitrous oxide by ammonia-oxidizing bacteria (see Sec. 5.2), provided ammonium is present. In aeration control, it is important to be aware of the risk of nitrous oxide emissions at low DO concentrations due to the accumulation of nitrite. However, reducing the DO concentrations with ammonia control when the ammonia load is low does not necessarily imply elevated concentrations of N_2O at all times. Stabilizing the process by providing the required oxygen for nitrification to occur can in some locations lead to a reduced risk of N_2O formation during ammonia peaks, since the aeration intensity is increased when this is necessary, avoiding low DO concentrations at periods of high ammonia load. If relevant, nitrite can be analysed to verify whether high nitrite concentrations occur. Reducing the risk of excessive aeration during periods of lower load also reduces the risk of oxygen escaping from aerated to non-aerated zones, which might disturb denitrification and lead to N_2O formation.

There are theories stating that there is an increased risk of sludge bulking at low substrate or oxygen concentrations (Martins *et al.*, 2004). The theory is that filamentous bacteria – responsible for sludge bulking – have easier access to substrate or oxygen thanks to their higher surface-to-volume ratio than do floc-forming bacteria, which means they are favoured during periods of low availability.

Consideration on ammonia control loops

There are limits on three controlled variables in an ammonia control loop: DO limits, airflow limits and valve opening limits. Which limit will be activated because of an increase or decrease of aeration intensity depends on local conditions such as turn-down capacity of the blowers, aeration capacity through the diffusers and selected DO limits.

If there are multiple zones with DO control with different limits it is not trivial to find which limit is first activated. Ammonia control provides the possibility to (1) limit aeration to save energy and (2) reduce effluent ammonia peaks (Rieger *et al.*, 2014). There are two main ways to control aeration: by changing the aeration intensity through varying the DO concentration or by changing the aerated volume by volume control.

In (1), aeration is limited by targeting an effluent ammonia concentration, which is greater than zero but with a margin to the effluent criteria. This is often best done with ammonia feedback control since the control authority for this controller is good when it comes to limiting aeration down to the specified low limit of the DO concentration.

For (2), ammonia feedback control has most often very little control authority. Nitrification capacity is determined by the mass of active nitrifiers in the system, which in turn depends on the ammonia load removed. Adding feedforward can do little to nitrification capacity during these situations, except when the hydraulic retention time of the system is short. If nitrification capacity is limited and DO concentrations are already around 2 mg/l everywhere in the process, the remaining option is to switch on an additional aerated zone to increase the mass of active nitrifiers.

Considerations on the use of PID controllers

For most applications, it is advised to use PI control for the feedback-loops in aeration control. PID controllers are well known and relatively easy to implement, there are few control parameters to tune and they perform well on the processes in an aeration tank. Using derivative action in the controllers is most often not necessary since the dynamics of the processes do not motivate more controller parameters to tune. Also, derivative control is sensitive to noise and signal filtering is required if used with noisy sensor signals – which are not uncommon in aeration processes.

5.3.3.2 Preliminary controller tuning

Tuning of airflow and DO controllers can be done using step responses and the lambda tuning method described in Chapter 4. Preliminary tuning in a model may be done but be aware that the controllers may be defined differently in the models and in the SCADA/DCS with different algorithms for calculating PID controllers and using different units in the controllers (e.g., engineering units or per cent).

Tuning an ammonia controller, including deciding on the ammonia setpoint, offers the operator a lot more options than when tuning the DO and airflow controllers. The settings should reflect the local conditions at the plant, where the most important are the effluent criteria and the aerobic sludge age, which could be important constraints as mentioned in the discussion on definition of goals in Sec. 5.3.

If the plant is required to limit the diurnal ammonia peak in the effluent, the focus on ammonia control is on disturbance rejection and energy efficiency is a secondary objective. A fast ammonia controller is best for disturbance rejection. Compared to many other applications, fast ammonia control is still slow given the long response time of the system itself. As emphasized earlier: in this cascade control system, the DO controller acting as a slave controller must be faster than the ammonia master controller. The integral time must be longer than that of DO control and reflect the response time of the aeration system including oxygen transfer. A typical fast controller may have an integral time of between one and two hours using a gain of around 1. This fast version of an ammonia controller requires an appropriate maximum DO limit in the controller to avoid energy loss.

At plants where a certain limit is never to be exceeded it is fruitful to try to push down the ammonia peaks since it allows for the plant to operate on a higher average ammonia concentration than without control (Ingildsen *et al.*, 2000). Finding the ammonia setpoint that provides enough margin to the effluent limit requires trial and error testing. This is one of the reasons why it is preferred to use a model for controller design since it can allow for extensive evaluation of the choice of ammonia setpoint.

In regions where the effluent criteria require the plant to reach an average effluent concentration the setpoint is not set to make sure the peaks never cross a certain limit. Instead the setpoint is chosen so that the average concentration is kept at a sufficient margin to the effluent criteria. If there is a criterion on total nitrogen concentration, nitrate control might also come into play to balance the total nitrogen removal (Ayesa *et al.*, 2006).

An average limit in the effluent criteria may allow for the use of slower ammonia control since disturbance rejection is not important. This can be implemented for instance using an averaging filter for the ammonia concentration (Ayesa *et al.*, 2006) or by a longer integral time or lower controller gain (Åmand *et al.*, 2014; Ingildsen *et al.*, 2000). Slow control can be used to find the best average DO concentration required to reach an ammonia setpoint over time. With daily ammonia peaks combined with the possibility to use low DO concentrations during low load periods (<1 mgDO/l) the energy saving will be lower with slower ammonia control since the controller may not reach low DO concentrations before the next peak occurs. If the ammonia controller is not allowed to lower the DO concentration below around 1 mg/l, there is not necessarily a large energy loss compared with fast ammonia control.

Measuring ammonia online with ion-selective sensors in the aeration tank can be challenging due to the hostile environment and disturbance from other ions. Remedies to this involve using an ammonia analyser with filtering or placing the ion-selective ammonia sensor at the outlet of the secondary settler. Both these remedies impose a time delay on the ammonia signal in relation to the actual concentration in the aeration tank. Time delays might call for a slower tuning of the ammonia feedback loop to avoid oscillations.

Note that the integral time is affected by the choice of controller gain if the controller is implemented on standard form (see Sec. 3). Another important note to make when tuning an ammonia controller is that the gain must be negative. If the effluent ammonia concentration is higher than the ammonia setpoint the ammonia control error will be negative and the DO concentration should be increased which implies a negative controller gain.

5.3.3.3 Decide on instrumentation and actuators

The control structure defines which variables to be controlled and which ones to manipulate, thereby specifying the required instrumentation and actuators in the process. Other important decisions involve how many sensors and actuators to use and where to place them. Is one ammonia sensor enough or should each treatment train be controlled separately? If there is a large difference in treatment performance each treatment train requires its own sensor for ammonia control to work satisfactorily.

The airflow requirement changes along a plug flow reactor when the pollutants are removed from the process. As mentioned in Sec. 5.3.3.1, a uniform airflow distribution will lead to severe DO peaks toward the end of the reactor and an attempt to calculate DO setpoints with the help of ammonia-based control would be fruitless. To allow for a balanced DO profile along a plug-flow reactor, DO sensors, airflow sensors and airflow valves should be in each compartment.

5.3.3.4 Develop safety nets and fallback strategies

The most important fallback strategies and safety nets in aeration control are the other cascade loops in the cascades and their limits. If an ammonia sensor freezes at a too low value, the DO concentration will end up at its lower setpoint. This is one of the reasons why airflow control is important, since it offers a meaningful safety net should the DO sensor fail.

Redundancy is another key for safety nets and fallback strategies. If a sensor fails and cannot be made to function directly, a possible fallback strategy could be to use a sensor placed at a similar position at a parallel treatment line, provided they show a similar behaviour. This can for instance be the case for ammonia sensors in parallel treatment trains or for DO sensors at identical compartments in parallel treatment trains.

5.3.4 Implementation

5.3.4.1 Upgrade instrumentation and actuators

Installation of new equipment is an important part of a successful implementation project, as described in Sec. 5.3.3. For instrumentation, it is suggested to follow these steps:

- installation
- configuration
- initialization
- base calibration.

5.3.4.2 Controller maintenance

Maintenance and supervision of controllers and control loops should be a part of the normal maintenance work at the plant, involving both preventive and corrective maintenance. If available, an industrial maintenance system should be used for supervision of the control loops, gathering data about controller parameters, performance measures and information about instruments and actuators.

Maintenance of controllers involves regular checks of the controller performance. Controller performance can be different things depending on the controller tasks. In the automatic control literature there is a range of more or less advanced methods for evaluation of controller performance. Here we just mention the most common options that are available: time constant of the closed-loop system and the standard deviation of the control error. If a controller, such as an airflow controller, is to track a changing setpoint, one of the most common metrics to use to check the controller performance is the *time constant* of the closed-loop system. The final tuning of controllers for the DO control is discussed below.

For disturbance rejection, the standard deviation of the control error is a common metric. Standard deviation can be helpful when, for instance, checking the settings of an ammonia controller dedicated to disturbance rejection.

One of the most common problems with controllers is oscillations. Oscillations are most often caused by: (1) poor tuning; (2) valves; (3) external disturbances (Olsson & Newell, 1999, chapters 17 & 24). A control loop may have oscillations due to one or several of these reasons.

There might be a need to change the controller settings due to seasonal changes such as temperature, variations in sludge age due to temperature variations etc.

In the maintenance plan for the controllers it is suggested that the following are included: (1) controller performance; (2) control loop behaviour including oscillations; (3) plan for seasonal changes to controller settings.

5.3.4.3 Poor control performance during operation

Even though a controller has been well tuned during commissioning, there will certainly come a day when the control performance starts to deteriorate. Typically, oscillating behaviour of the process variable is observed. The instinct may then be to directly try to retune the controller (and blame the controller!). Often, however, it is not a controller issue. The following check is recommended before retuning the controller:

- Put the controller in manual mode. If the oscillations persist it is not the controller to blame. Then check for disturbances 'upstream'. For example, a control system in a previous step may have gone unstable.
- If the oscillations disappear when the controller is put to manual mode it is recommended to first check the actuators and the sensors before a retuning is done. A typical actuator fault is an error in the valve such as increased friction. These types of problems may then be the cause for the poor control performance.

5.3.4.4 Sensor and actuator maintenance

Care should be used with all sensors. Without a clear commitment to maintain them properly, they may eventually become useless (at best) or cause more harm than good. A rule of thumb that could be employed is

$$Sensor\,maintenance \sim 30-60\,min/week/sensor$$

for general maintenance. For a large plant this would result in one or more full-time operator positions, which would be likely considered unsustainable. Hence, the need for automation and intelligent systems to detect faults and prioritize maintenance options. In a sense, we can think of this approach as a *sensor triage*.

Imagine the case when a sensor is completely fouled and does not give a realistic reading, triggering a control system to respond in ways that are clearly not responsive to the process needs. In this case, which occurs surprisingly often, the aeration system may be experiencing turn on/off events at peak power charge times, associated with stiff penalties and with large sums of money wasted. An additional, yet less clearly understood, potential pitfall of probe neglect is the over-aeration of the biological reactors, with the associated excessive shear imparted by the aeration system to the biomass and the consequent sporadic (if not systematic) upsets of the secondary clarifiers. The same may happen when the process is operated at an excessively elevated DO, due to equipment limitations, overzealous operations, or simply *laissez-faire* lack of interest in the process.

Recent work by the authors has been demonstrating the importance of understanding the value of good sensor maintenance and the potential dangers of overzealous sensor care. In fact, excessively frequent sensor re-calibration and signal compensation may lead to error and uncertainty (Cecconi *et al.*, 2018). Therefore, the instructions from the technology developers of sensors and actuators should be diligently followed. After all, they should know their equipment better than anybody else.

Another important element in the healthy conduct of aeration operations is the regular exercising of actuators. Many elements of the aeration system are designed for variable operations but are *de facto* operating always in the same position (almost always these are in their normally open or closed position). Each actuator should be exercised regularly, to avoid the effect empirically known as 'sticky valves'.

5.3.4.5 Implementation of control solution in control system

There is a range of different industrial control systems and SCADA systems running at WRRFs around the world. Some plants have clear yet flexible implementation of the treatment processes for monitoring and control, and a new or altered control solution for aeration control might not be much of a challenge to add. However, this is not always the case and some suggestions to consider in relation to aeration control are given below.

During the implementation process always make sure that lower-level controllers work well before adding extra controllers on top of the existing ones. Work through the blower control, pressure control, airflow control, DO control and ammonia control loops in this order and see to sensor signal quality, actuator performance and controller tuning. The performance of the outer control loops depends on the performance of the inner control loops especially when the outer controllers calculates setpoints to the inner controllers as is the case in cascade control.

PID controllers have been installed in control systems in process industries for decades. Nevertheless, there are aspects of implementation that are sometimes overlooked, some of them being very important in aeration control. Regardless of who is actually implementing the controller, we present here some points that we highly recommend.

When implementing the PI-controller parameters, there are a couple of things to think about in relation to aeration control.

- The controller gain, K, can be negative. In DO and airflow control the gain is positive, but in ammonia control the gain should be negative since a higher DO concentration leads to a lower ammonia concentration.
- The integral time should be allowed to be long when dealing with ammonia control due to the long response time. Be advised not to choose too few

significant figures in the interface when implementing PI control. Sampling rate of the air supply has already been mentioned.

- Always make sure that a control signal has a high and a low limit. This is the simplest form of safety net should a sensor malfunction, and is crucial to avoid unwanted drift of the process. Limits for the most common control structures are defined in the design procedure suggested above. Should there be any other solutions implemented in the control system, where for instance two control signals are combined to one, all these three signals must have limits.

Cascade control is a cornerstone in state-of-the-art of aeration control. There are two special cases about implementation of cascade controllers:

(1) avoid jumps in slave control signal when a slave controller shifts from having an internal setpoint (given by operator) to external setpoint (calculated by master controller);
(2) avoid windup of the master controller when the slave controller saturates, for example, reaches the operation limits, as described in Sec. 5.3.3.

The solution to (1) is to let the master controller track the setpoint of the slave controller, allowing for a smooth transition when the master controller setpoint is to take over from the manual setpoint. For (2), the master controller needs to track the control signal and its limits and inhibit integral action in the master controller when a limit is reached (read more in Sec. 5.3.3). Many industrial control systems on the market handle these special cases well, but it is never incorrect to ask the control system vendor or programmer to be sure.

Another aspect of implementation of cascade control is transparency. Large plants may have multiple parallel treatment trains, perhaps with several compartments with DO and airflow rate control and perhaps ammonia control on top. In these situations, make sure the operators can have a good overview of the different cascades and that it is clear in the user interface which signal is an output from one controller, and an input to another. An example of a transparent implementation of cascade control is found below. During tuning of the controllers, a transparent solution during implementation can make all the difference.

5.3.4.6 Aeration control case study

At Kungsängen WWRF in the city of Västerås, Sweden, the aeration controllers are implemented with a standard cascade control structure, see Section 3. The DO and airflow controllers are implemented in the control system so that the communication between the cascade controllers is clearly visible, see Figure 5.2. This allows for simpler tuning and maintenance of the controllers.

The plant control system is ABB 800xA. The plant has experienced personnel within the organization who are experts in control system programming in general and in programming ABB systems in particular.

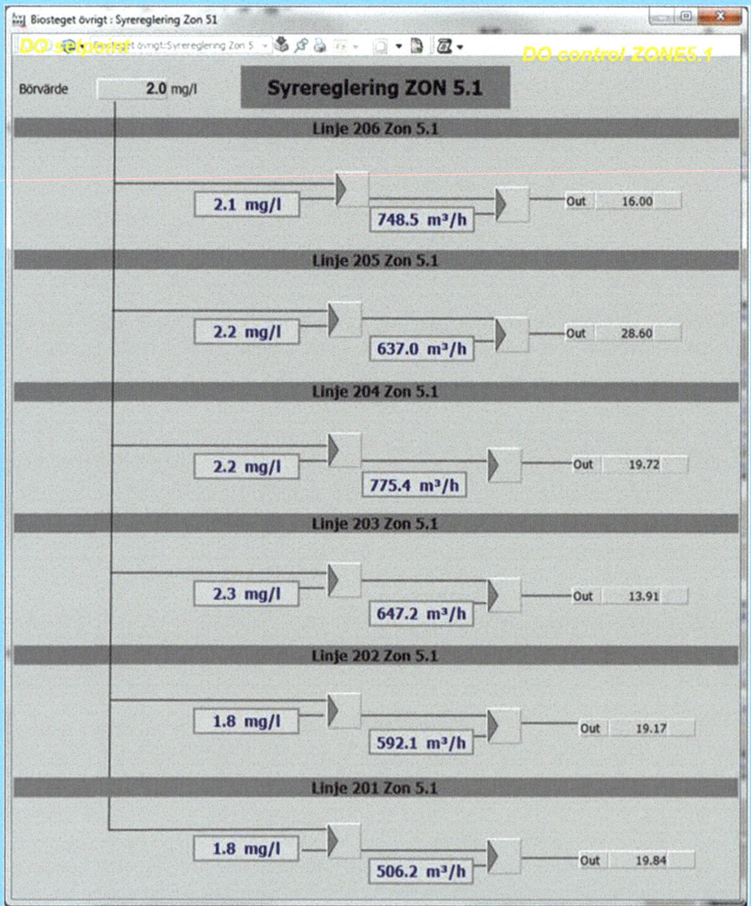

Figure 5.2 The DO and airflow rate controllers are implemented in a clear and transparent fashion in ABB 800xA at Kungsängen WWTP. In this example the controllers at the same zone (zone 5.1) in six parallel treatment trains (trains 201 to 206) are shown. The measurements of DO and airflow rate are displayed in blue as inputs to each controller, together with the setpoints. 'Out' refers to the control signal to the valves. The possible drawback of this example is that all zones have the same DO setpoint, but that can relatively easily be changed with this type of implementation.

5.3.4.7 Final controller tuning

If models are used during the design phase, estimations of controller parameters are available. The final controller tuning is done after the controllers are implemented in the control system. If there was no tuning made in a model, all controller tuning is done during implementation. Controller tuning using the lambda method or other methods based on step responses requires that the input signal and control signal in a controller can be tracked with sufficient resolution. To record a step response of an airflow controller only every 30 seconds is not sufficient. An example of lambda tuning of an airflow controller is found below.

PID controllers can be presented mathematically in different forms, and the standard form (see Sec. 4.3) is the most common in industrial applications. If model-based controller tuning was carried out during the controller design phase the estimated controller parameters should be implemented in the real process, and these must be re-calculated if the controllers in the model and at the plant are not in the same form or if the model and the control system use different units in the calculations.

After implementation of actuators, sensors, controllers and final tuning of each control loop, the next step involves evaluation of the full control solution and how the different controllers interact, which may require a few weeks of performance monitoring.

5.3.4.8 Example of lambda tuning of airflow controller

Lambda tuning was described in Chapter 4. Two step-response tests were performed at different influent loads during the day, as illustrated in Figure 5.3. The lambda tuning method was used to calculate K and T_i in the

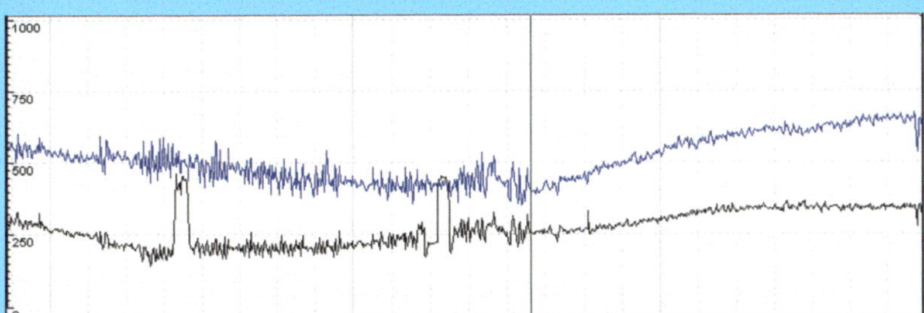

Figure 5.3 Demonstration of lambda tuning. Black curve, step-response test in aerated zone; blue curve, use of these new settings in adjacent aerated zone. The left and the right parts of the figure show two different tuning events. Data from 24 hours, unit on *y*-axis: N m^3/h.

PI controllers. At the marking the new settings were used. The two step responses resulted in similar response times. In the lambda tuning a *p*-value of 3 was used.

Changing to slower control settings ($T_i = 15$ seconds) after lambda tuning resulted in a smoother control of the airflow rate. The step-response test was performed in one of the aerated zones (black line). The controller parameters were changed to the same new values in an adjacent aerated zone (blue line), which improved the performance in a similar manner.

5.3.5 Operation and maintenance

5.3.5.1 Maintenance of controllers, actuators and instrumentation

It is now time to execute the plans for maintenance of controllers and equipment developed during the implementation phase. With a well designed control solution, implemented and tuned this should be a straightforward task.

A final reminder on cascade control is that control loops in cascade control should not be maintained as separate units but as parts of a larger control solution. Always tune the inner control loop first. An example of what might happen if this is not done is described here.

5.3.5.2 Oscillations in DO cascade control

Since an outer control loop calculates the setpoint to an inner control loop the control loops interact with each other. To avoid an unstable system, the inner control loop must always be faster than the outer loop. Oscillations can also

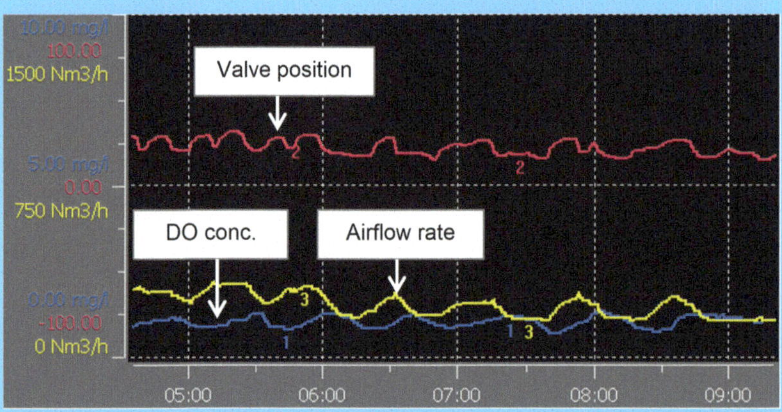

Figure 5.4 Oscillations in a DO cascade control loop with airflow control. The oscillations were induced by the too short integration time (T_i) in the DO controller (60 s), which after an adjustment to 300 s solved the problem.

occur even though this rule is followed, but if the controllers are poorly tuned. This may happen over time if the controllers are not considered as a unit when controller parameters are changed.

An example of such behaviour is presented in Figure 5.4. The oscillations were more severe at night when the load to the plant was lower. The poorly tuned controller caused oscillations in the DO sensor signal of up to 0.5 mg/l in peak-to-peak amplitude.

5.3.5.3 Updates and improvements

Despite a successful implementation, changes and updates in all systems are to be expected, especially during the initial period of operation.

5.3.5.4 Operator training and education

Operator training and education is a cornerstone for successful operation of a new control solution. Already at the end of the implementation phase it is important that personnel working with the biological treatment steps know about performance evaluation and learn how to perform and document maintenance of controllers, sensors, and actuators.

5.3.5.5 Develop and calculate performance metrics

The performance metrics should reflect the original objectives of the project. Examples of possible performance metrics are listed in Table 5.4, with benefits and possible drawbacks.

The wastewater operation is periodic with daily cycles, weekly cycles and annual cycles. Performance of a new aeration control solution is best compared on an annual basis. A control solution is likely to have a larger impact on the plant in some parts of the year, for instance during the warmer period when there is a larger potential to decrease oxygen concentrations. Evaluating a full year of operation means temperature and load changes are considered in the evaluation and an over- or under-estimation of the annual benefits may be avoided. This may allow for a better cost/benefit analysis.

Having good-quality data for several years in a row (before and after implementation) is at a minimum challenging at most treatment plants. Perhaps instrumentation was not there all the time or moved between different locations, treatment trains were taken out of operation for maintenance work, airflow meters did not function correctly part of the time, etc. At times, the modification of the sensor position or the abandonment of a control system are dictated by the convenience to operators and are not registered or documented until an issue reveals them.

An alternative to annual comparisons of metrics is to have dedicated measurement campaigns before and after implementation of a new control

Table 5.4 Examples of performance metrics for evaluation of a new aeration control solution.

Performance Metric	Suggested Unit(s)	Comment/benefit	Possible Drawback
Cost/benefit analysis		Includes annual operation costs such as maintenance costs and compares this to the benefits of the new solution	Challenging to estimate all the benefits, such as reduced need for manual operations
Payback time	Years	Similar to above	Similar to above
Operation costs	Monetary unit	Includes costs other than aeration, such as pumping, chemicals and maintenance costs	Risk of bias due to difference in load and removal rates during evaluation periods
Energy cost for aeration	Monetary unit	The actual cost of operating the blowers, including possible effects of the energy tariff.	Risk of bias due to difference in load and removal rates during evaluation periods
Power demand	kWh	The actual energy demand for aeration. If the energy tariff is flat, this is comparable to Energy cost	Risk of bias due to difference in load and removal rates
Average airflow rate	N m^3/d, N m^3/yr	Comparable to power demand only if there is a linear relationship between airflow rate and power demand.	Total airflow rates are often correct but the flow meters at individual process trains cannot always be trusted. Does not consider the possible effect of energy tariff.

Specific energy consumption or cost	N m^3 or kWh or energy cost per unit load removed (e.g., kgCOD, kgBOD, $kgNH_4$)	Reduce the bias of the energy metrics by normalising with the actual amount of nitrified ammonia.	Specific energy consumption goes up when the load goes down and less ammonia is removed. Hence, the metric is still biased. Requires measurements of incoming and outgoing ammonia to the aeration process. May or may not include the organic load. When organic load is included, its measurement may be delayed (e.g., for BOD) or indirect (e.g., COD)
Effluent concentrations	mgN/L	Average NH_4, NO_3 and TN concentrations. Daily, monthly and/or annual average depending on effluent criteria.	Effluent monitoring may be performed intermittently or only at fixed times each weekday, to meet regulatory requirements
Number of effluent violations		Important when grab sampling used for test of compliance with effluent criteria	Grab samples may be gathered in non-representative times or locations
Variation of effluent concentrations	$(mgN/L)^2$ or mgN/L	Variance or standard deviation of effluent NH_4, NO_3 and TN.	

solution. This is an option that allows for good-quality data for a shorter period, but this method may not offer an evaluation of the actual performance at the plant over time since seasonal variation is not taken into consideration. The seasonal variation may play a more important role at plants where temperature and load vary more, such as locations with snow and ice melting periods or seasonal tourist periods.

Comparisons can also be made between two parallel treatment trains. The benefit of this method is that systematic differences between different evaluation periods are not an issue. Instead, one must be on the lookout for systematic differences between the treatment trains. These may be substantial and airflow requirements can differ by 5 to 10%. Not all plants measure the flow rates to the individual treatment trains.

No matter which method is chosen to evaluate the performance of the aeration control solution, always be aware of possible systematic differences in the data, as summarized in Figure 5.5.

Are changes in effluent concentrations or energy consumption due to a change in control solution significant or not? Use statistical testing (such as the Student's t-test) to find out whether two data sets are statistically significantly different from each other.

5.3.5.6 Compare to objectives and criteria

Did the new aeration control solution meet the objectives? Were all the criteria met? Were the performance metrics useful? Often changing the control solution results in more questions than those asked at the start of the project. If there is a need, an update of the objectives and a changed implementation can be considered. Otherwise, a successful implementation project will move on to become a new part of the everyday operations and maintenance of the treatment plant.

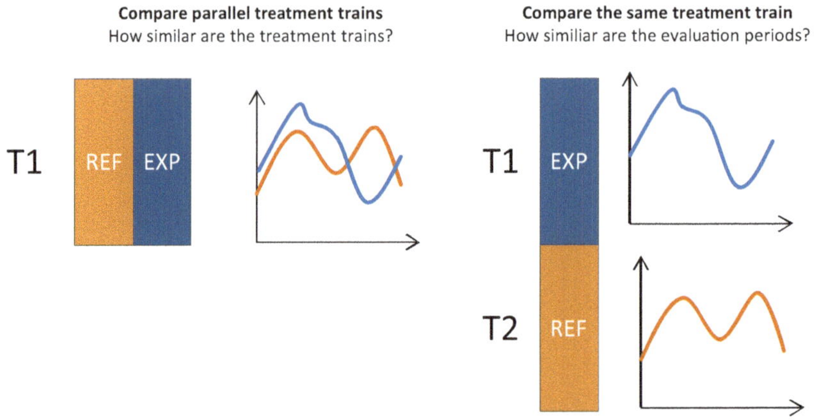

Figure 5.5 Questions to ask when evaluating the difference between two control solutions. Adapted from Åmand *et al.* (2014).

5.4 OUTLOOK

Aeration control is well established, and is arguably understood better than practiced in most facilities. Three categories of development make it a proven and successful technology:

- sensors for aeration control are reliable and affordable;
- control systems are well developed, and computer implementations are readily available;
- flexible actuators are available, and variable speed drives are standard nowadays, making blower/compressor control flexible.

Experiences from many cases of full-scale implementation can be of guidance for future installations.

5.5 REFERENCES

Åmand L., Olsson G. and Carlsson B. (2013). Aeration control – a review. *Water Science and Technology*, **67**(11), 2374–2398.

Åmand L., Laurell C., Stark-Fujii K., Thunberg A. and Carlsson B. (2014). Lessons learnt from evaluating full-scale ammonium feedback control in three large wastewater treatment plants. *Water Science and Technology*, **69**, 1573–1580.

Ayesa E., De La Sota A., Grau P., Sagarna J. M., Salterain A. and Suescun J. (2006). Supervisory control strategies for the new WWTP of Galindo-Bilbao: the long run from the conceptual design to the full-scale experimental validation. *Water Science and Technology*, **53**(4–5), 193–201.

Cecconi F., Reifsnyder S., Donikian N. A., Ito Y., Jimenez M., Sobhani R. and Rosso D. (2018). Suitability of High/Low range ISE-ammonium sensors for monitoring and control purposes in WRRFs. In: *Proceedings of IWA EcoSTP Conference*, London, Ontario, Canada, 25–27 June 2018.

Foley J., de Haas D., Yuan Z. and Lant P. (2010). Nitrous oxide generation in full-scale biological nutrient removal wastewater treatment plants. *Water Research*, **44**, 831–844.

Funk W., Dammann V. and Donnevert G. (2007). *Quality Assurance in Analytical Chemistry*, 2nd edn. VCH, Weinheim, Germany.

Ingildsen P. (2002). Realising full-scale control in wastewater treatment systems using in situ nutrient sensors. *PhD Thesis*, Lund Univ., Sweden. Available online at: www.iea.lth.se/publications/Theses/LTH-IEA-1030.pdf (accessed 10 Feb. 2018).

Ingildsen P., Olsson G. and Jeppsson U. (2000). Using nutrient sensors for estimation and control of a full scale activated sludge process. In: *Proceedings of WaterTech Enviro 2000*, Sydney, 9–13 April 2000.

ISO 15839 (2003). *Water quality—online sensors/analysing equipment for water—specifications and performance tests*. ISO 15839, Geneva, Switzerland.

Kampschreur M. J., Temmink H., Kleerebezem R., Jetten M. S. M. and van Loosdrecht M. C. M. (2009). Nitrous oxide emission during wastewater treatment. *Water Research*, **43**, 4093–4103.

Martins A. M. P., Heijnen J. J. and van Loosdrecht M. C. M. (2004). Bulking sludge in biological nutrient removal systems. *Biotechnology and Bioengineering*, **86**, 125–135.

Montgomery D. C. (2009). *Introduction to Statistical Quality Control*, 6th edn. John Wiley, New York, NY, USA.

Olsson G. (1985). Dynamics and control of wastewater treatment systems. In: *Final address, 4th IAWPRC Workshop on Instrumentation and Control of Water and Wastewater Treatment and Transport Systems*, Denver, CO, USA, May 3, 1985.

Olsson G. (2012). ICA and me – a subjective review. *Water Research*, **46**, 1585–1624.

Olsson G. and Newell B. (1999). *Wastewater Treatment Systems*. Modelling, Diagnosis and Control, IWA Publishing, London, UK.

Olsson G., Nielsen M., Yuan Z., Lynggaard-Jensen A. and Steyer J. P. (2005). *Instrumentation, Control and Automation in Wastewater Systems*. IWA Publishing, London, UK.

Rieger L., Siegrist H., Winkler S., Saracevic E., Votava R. and Nadler J. (2002). In-situ measurement of ammonium and nitrate in the activated sludge process. *Water Science and Technology*, **45**(4–5), 93–100.

Rieger L., Langergraber G., Thomann M., Fleischmann N. and Siegrist H. (2004). Spectral in-situ analysis of NO_2, NO_3, COD, DOC and TSS in the effluent of a WWTP. *Water Science and Technology*, **50**(11), 143–152.

Rieger L., Thomann M., Gujer W. and Siegrist H. (2005). Quantifying the uncertainty of online sensors at WWTPs during field operation. *Water Research*, **39**, 5162–5174.

Rieger L., Gillot S., Langergraber G., Ohtsuki T., Shaw A., Takács I. and Winkler S. (2012). *Guidelines for using Activated Sludge Models. Scientific and Technical Report No. 22.* IWA Task Group on Good Modelling Practice. IWA Publishing, London, UK.

Rieger L., Jones R. M., Dold P. L. and Bott C. B. (2014). Ammonia-based feedforward and feedback aeration control in activated sludge processes. *Water Environment Research*, **86**, 63–73.

Stare A., Vrečko D., Hvala N. and Strmcnik S. (2007). Comparison of control strategies for nitrogen removal in an activated sludge process in terms of operating costs: a simulation study. *Water Research*, **41**, 2004–14.

WEF (2013). Automation of Water Resource Recovery Facilities, 4th edn. *Manual of Practice No. 21*, Water Environment Federation, Alexandria, Virginia, USA.

Chapter 6

Energy intensity of aeration

Diego Rosso and Manel Garrido-Baserba
University of California, Irvine

6.1 THE ROLE OF AERATION ENERGY

Conventional wastewater treatment plants (WWTPs), or as they have been renamed recently, water resource recovery facilities (WRRFs), as their role as sources of energy and materials is being more fully recognised, need to embrace a whole set of complex processes and operations, many of them characterized by high energy requirements. The complexity of transforming urban wastewater into 'clean' water, or at a minimum, into 'suitable' water before its reintroduction into the environment, involves the removal, transformation, and handling of an almost infinite number of pollutants and process by-products, which will imply the coordination of dozens of physical, chemical and biological operations, all of them energy-dependent. The combination of all these operations undoubtedly will require a high amount of electrical power, converting WRRFs into one of the largest energy consumers in developed countries. Approximately 2–3% of the world's electrical energy is used for water supply and sanitation purposes, and 1–18% of the electrical energy in urban areas is used to treat and transport water and wastewater (Ingildsen & Olsson, 2015; Olsson, 2012). Hence, understanding all the energy consideration in WRRFs is of the upmost importance to optimize, design, and implement the most convenient energy-reduction measures.

Editor: Diego Rosso
doi: 10.2166/9781780407845_0179

The energy requirements for these operations will be dictated by their own technological features and characteristics. Different technologies or equipment can be used for the different processes within the WRRF process flow diagram (i.e., disinfection, settling…), albeit yielding different energy consumption. Variables ranging from reliability, control capacity, efficiency, maintenance, etc., may lead to different energy demands for the same operation. The energy intensity, however, of treating wastewater is dependent less on the type of technologies installed than on both the level of treatment (quality targets) and the size of the plant or water volume (economy of scale). Depending on the pollutants to be removed and the extent of removal efficiency pursued, the energy demand will largely vary (Table 6.1). Three main types of treatment quality targets with their corresponding energy intensities are usually considered. The first one, known as primary or mechanical treatment, focuses exclusively on the partial removal of carbonaceous matter (COD, BOD_5) and total suspended solids by sedimentation or filters, and the solids are removed by scrapes. The energy intensity of primary treatment is obviously very low in comparison to more extensive treatment using biological processes. Biological treatment (which will include always the primary treatment) will increase energy intensity due largely to one process: aeration. Aeration involves helping the microorganisms to 'breathe' by supplying oxygen in the form of dissolved oxygen, so they can feed with the nutrients (pollutants) found in the wastewater.

Depending on the extent of the aeration, two different main types of operation can be considered. While the removal of organic matter alone requires less oxygen per amount of wastewater, the removal of other nutrients (i.e., nitrogen) sets a high aeration energy demand with the concurrent increase in energy requirements. While only 1.2 g of oxygen are required to degrade 1 g of BOD, more than 4.5 g of oxygen are needed for full nitrification (<2 mg/l NH_4^+).

Energy for wastewater treatment has to relate to the pollutants removed (or level of treatment), although the energy intensity of treating wastewater also depends on the size of the plant (Table 6.2).

A sharp increase in energy intensity is also expected when it is necessary to improve water quality by a more extensive removal of pollutants of potential concern. So-called tertiary treatment, which may involve the reduction of

Table 6.1 Energy requirement for wastewater treatment in Australia.

Type of Operation	kWh/m³ (min–max)	kWh/m³ (average)
Primary	0.1–0.37	0.22
Biological C removal (incl. primary)	0.26–0.82	0.46
Advanced C, N and P removal	0.39–11	0.90

Sources: Kenway *et al.* (2011); Sydney Water; Brisbane Water (Kenway *et al.*, 2011).

Table 6.2 Energy requirement for wastewater treatment in the US.

Plant Size		Electrical Energy Consumption	
m^3/s	MGD	Activated Sludge Biological C Removal	Activated Sludge with Nitrification
0.044	1	0.59–0.69	0.78
0.44	10	0.32–0.37	0.47
4.38	100	0.27–0.31	0.41

Sources: Griffiths-Sattenspiel *et al.* (2009).

emerging contaminants of concern, refractory substances and pathogenic organisms in general, may imply a remarkable increase in the overall energy requirements. Processes enabling the treatment of wastewater to drinking or close-to-drinking quality, like advanced filtration (i.e., reverse osmosis, nanofiltration, ultrafiltration…) will further raise the energy intensity.

6.1.1 Aeration in biological wastewater treatment

Aeration is a critical operation in most wastewater treatment processes. Among all required operations in water resource reclamation facilities (WRRF), pumping air into the biomass (by blowers or compressors) is recognized as one of the (if not ***the***) most energy-hungry process in wastewater treatment. Air needs to be continuously supplied to the wastewater to guarantee an adequate concentration of dissolved oxygen (DO) to promote microbial growth. DO concentration governs the removal of the main pollutants of concern: carbonaceous matter and nitrogen. The rate of removal will increase with the dissolved oxygen concentration, although to a certain limit. Aeration can be achieved with mechanical or diffused systems, and the latter may be with either coarse- or fine-bubble diffusers. As seen in Chapter 2, fine bubble diffusers are capable of transferring 2–3 kg O_2/kWh, while coarse bubble diffusers deliver only about 1 kg O_2/kWh.

Adding or increasing the dissolved oxygen concentration in wastewater is a very inefficient process that requires a constant supply of enormous volumes of air through the corresponding system of diffusers to satisfy the DO requirements. Energy-demanding blowers (or compressors) deliver air through the diffusers installed in the floor of the aeration basin. The bubble plume from the bottom of the tank to the surface are responsible for transferring dissolved oxygen to the water. Aeration basins typically have an approximately depth of 4–6 m, although deeper tanks exist and require the use of compressors (due to elevated hydrostatic pressure). The limited length of the water column limits oxygen transfer.

A short contact time for oxygen mass transfer is not the only constraint in aeration. Process inefficiencies are key to understanding the high energy intensity in wastewater aeration. On one hand, oxygen transfer via bubble columns is extremely limited (the rule of thumb is ~2%/ft or ~6.6%/m of oxygen is transferred in clean water). Moreover, because blowers supply air, not pure oxygen, only 21% of the total volume is actually oxygen, resulting in massive quantities of air being blown through the aeration system for the sole purpose of mixing. This low efficiency is hampered further by the effect of certain contaminants (i.e., surfactants). The so-called surfactants (or surface-active agents) from soaps and detergents immediately cover the bubbles' surface, hindering the mass transfer between gas and liquid phase. Oxygen mass transfer reductions to less than 0.40 of the clean water oxygen transfer can be typically expected (for alpha factors, see Chapters 2 and 7). The cumulative effect of both inefficiencies and depth constraints due to hydrostatic pressure results in a poorly efficient process. Only roughly 2–4% of the total supplied air volume can be transferred into the wastewater as dissolved oxygen.

6.1.1.1 Energy requirements in the aeration process

The electrical energy consumed by the blowers to guarantee that microorganism are adequately supplied with required DO at all times can account for more than 50% of the net power demand (Reardon, 1995; Rosso & Stenstrom, 2005). Energy consumption as high as 75% has been reported (Rosso *et al.*, 2008). These results are in accordance with an energy baseline benchmarking study that measured energy consumption in the secondary treatment processes of various sizes of plants, confirming that energy consumption of secondary treatment ranged from 30 to 57% of total plant energy consumption. As previously shown, the energy consumption is strongly influenced by the effluent quality demanded. Where only carbon removal is applied, the aeration requirements represent about 40% of the total energy demand. Energy consumption will increase to approximately 50% if biological nitrification/ denitrification is applied, and up to 65–70% in the case of enhanced nitrogen removal. Hence, there is an evident trade-off between the effluent quality reached and the energy investment and environmental footprint (Guo *et al.*, 2016).

Values consistent with those presented in Figure 6.1 are extensively reported worldwide. Therefore, aeration has become the main focus of considerations to reduce energy use in the ultimate aspirational goal of achieving net energy neutrality for water resource recovery. As an example, we can see the energy consumption of WRRFs ranges from 335 MWh month^{-1} (WRRFs serving 100,000 population-equivalent, or PE) up to 6600 MWh month^{-1} (WRRFs serving 3,000,000 PE), while the associated energy costs can range from 45,000 € month^{-1} to 280,000 € month^{-1}, respectively.

Such requirements are consistently shown as the major operational component in the present value calculation of treatment costs spanning from 30 to 45% (Kaliman *et al.*, 2008; Reardon, 1995; Sobhani, 2011; Sobhani *et al.*, 2011; WEF, 2009). Due to the elevated energy intensity, it is of the utmost importance to understand aeration

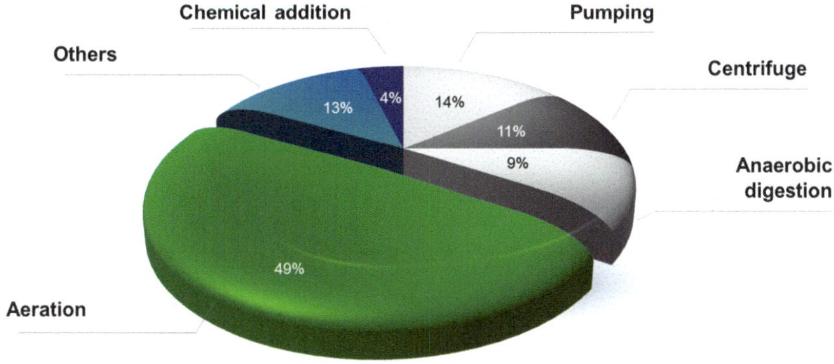

Figure 6.1 Estimated power usage breakdown for a typical 20MGD (74,500 m^3/day). Figure adapted from WEF (2009).

efficiency in order to improve design, performance, costs predictions, and process sustainability. Understanding aeration efficiency and lowering operating costs go hand-in-hand (Rosso & Shaw, 2015; Rosso & Stenstrom, 2006a).

6.2 POWER DEMAND, ENERGY CONSUMPTION, AND TARIFFS

In order to understand how process dynamics should be considered (or reconsidered), in our era of consciousness about the environmental and economic impacts of energy importation, we should spend some time entering the energy arena (that is why *sparks* is a necessary key word in this book's title). Perhaps when most treatment facilities in the western world were developed, little or no consideration was given to energy costs and wastage. We can imagine that at a time when energy costs were considered irrelevant, the energy supply was considered a given and virtually free component of the infrastructural design, just as air is free as far as our blowers are concerned. The reality nowadays is that many a plant is planned for greenway construction in emerging economies areas worldwide, and we have the opportunity to front the limitation in energy resources with more efficient design and operations.

WRRFs are large energy consumers if we look at their individual power bill, but are a small fraction of the societal energy footprint overall. In fact, the energy *per capita* we invest in transportation, heating, etc., is far larger than the treatment and reclamation of resources from wastewater (Olsson *et al.*, 2014). The large power bill of WRRFs is mainly due to their influent pumping and the energy intensity of the aeration process. Whereas pumping is a well understood operation, much is left to be understood about aeration, leaving it as an interesting field of research and of (sometimes challenging) practice. The large energy demand to satisfy DO requirements translates into large associated costs. Understanding the real cost of the aeration process is key in WRRF management. The implementation of measures to reduce peak power demand can contribute to power grid stability, decreased CO_2 emission, and the reduction of the energy bill.

Power demand and energy consumption are priced following a set of structures established to promote a rational use of the existing infrastructure and energy sources. Such pricing structures (also called energy demand-side mechanisms) are used to improve the energy system in terms of consumption and stability. Through the application of different energy pricing structures (e.g., time-of-use rates) and charges (e.g., energy usage, peak power demand charges) in the different billing terms, energy generation and distribution is optimized to reduce or shift peak power demands. Investments in additional infrastructures are avoided by balancing energy use and reducing greenhouse gas (GHG) emissions.

However, those savings-oriented strategies leave WRRFs with little room for manoeuvre, whose daily loading dynamics often clash head-on with established energy tariff structures from the power utilities. WRRFs typically receive the highest loading flowrate (to be treated) when the cost of energy is also the highest. The consequent overlap between the receiving influent peak and the most expensive energy price exacerbates both the cost of treatment and the concurrent GHG emissions. Even small shifts in peak demand have a large effect on savings to consumers and avoided costs for additional peak capacity. A study concluded that a 1% shift in peak power demand would result in savings of 3.9% – billions of dollars at the system level (Spees & Lave, 2008). Such large reductions would be achieved after encouraging customers and industries to properly adjust their energy consumption and reduce peak power demands.

Using an example from the Southern California Edison (SCE) tariff structure, it was found that electricity prices can be almost 25% more expensive depending on the time of use, in a plant with a ratio of 4.2-fold (for airflow rate and energy consumption) from peak to minimum (Emami *et al.*, 2018). For this specific site it was found that the maximum electricity consumption (160 kWh) occurred during the mid-peak period rates ($0.071/kWh) while falling to a trough of electricity consumption (37.9 kWh) during off-peak rates ($0.056/kWh). It was also expected that once power demand charges were included, energy costs were actually further amplified, confirming how energy tariff structures can drastically affect aeration energy costs.

6.2.1 Peak shaving strategies

Several potential measures can be applied to reduce or shift the power demand of WRRFs. The concept of peak shaving always has the same goal, to promote the shifting of the most energy-intensive periods of treatment to those when energy is cheaper and carbon-associated emissions lower. Three main types of strategies are currently available. However, with the current pace of energy renewables and new energy storing systems coming into the picture, the energy landscape is experiencing fast changes.

6.2.1.1 Energy conservation

Equalization tanks

The most traditional methods for peak shaving involve storing the wastewater in specific basins so it is possible to control when to treat water with more flexibility, preferably at night when energy is cheaper. The strategy to install flow and load equalization tanks was evaluated as a strategy to shift power demand by Leu *et al.* (2009) for a case study in California. Their results showed decreased costs and even reduction of CO_2 emissions at the energy generation side. The energy consumption reduction will depend on the investment capacity to build equalization tanks, the specific tariff structures at the site, and WRRFs and influent load characteristics.

Aeration control

Control of aeration has been successfully brought into practice with reductions in energy consumption as high as 30% (Åmand *et al.*, 2013; Olsson, 2012). These reductions have been converted into monetary units by using an average energy price (Benedetti *et al.*, 2009; Cadet *et al.*, 2004; Ekman *et al.*, 2006; Guerrero *et al.*, 2012; Samuelsson *et al.*, 2007; Stare *et al.*, 2007) or non-monetary units using the Operational Cost Index (Gernaey *et al.*, 2014). However, little attention has been given to energy tariff structures with respect to the evaluation of control strategies or technologies in view of energy cost.

Energy storage

The emergence of both renewable energies and new concepts in energy storage has introduced new tools to shave energy peaks in WRRFs. The steady increase and incorporation of renewable energies within the electrical mix is leading to a reduction of the associated carbon emissions in energy peaks hours (kg_{CO2}/kWh). For example, in the case of California where the power demand peak is considered to be at midday, the introduction of solar energy is helping to reduce the production and importation of fossil-based energy, reducing the associated carbon emissions. The impact is such that energy suppliers are quickly taking measures to address the postponement of the traditionally considered 'peak hours' (where energy is more expensive) happening at midday to the evening/night-time.

While solar and wind energy are reshaping current tariff and emission schemes, storage alternatives are bringing new concepts to help rethink the WRRFs' capacity to shave energy peaks. The introduction of energy batteries with the objective to store energy is becoming a reality in some WRRFs. Batteries can store power during off-peak times and then discharge it to reduce peaks, all without interrupting operations at these must-run facilities. For example, a current large project for a water district included the installation of lithium-ion batteries at eleven different sites across the Southern California region and used them to reduce their draw on the grid when the power utility needed. The strategy also included a special arrangement with the water district, whereby the energy storage company offered a better price in exchange for installing more batteries than required, so that the company can use them as surplus capacity for other purposes. This win–win situation enabled to save money in the acquisition of space somewhere else by the storage provider, while the treatment facility benefited from cheaper energy storage prices just for allowing them to use unused space.

WRRFs are slowly taking advantage of the carbon-rich influent and transforming the treated sludge into energy-rich biogas using technologies such as anaerobic digestion. Anaerobic digesters allow the production of biogas, which can be conveniently used to help abating power demand peaks. Proper storage systems together with combined heat and power (CHP) technologies are helping to reduce peak hours by using their own produced biogas, avoiding the acquisition of expensive and high-emissions-associated energy. The most common ones are internal combustion engines (ICE), while micro-turbines are quickly entering into WRRFs. As of 2011, according to a US EPA Combined Heat & Power Partnership report, 50% of US WRRFs with CHP used ICE technology. ICE generators can operate in several different modes to provide electricity to the water treatment facility and, in some cases, to the electric utility grid. Options include 'base load', 'peak shaving', and 'island' modes, where all generated power is consumed 'behind the meter' by the water treatment facility. Island mode means that the ICE equipment is isolated from the electric utility and serves specific on-site loads. 'Base load' and 'peak shaving' are usually most effective when operating in parallel with the electric utility. However, electric utility interconnection requirements may lead to impractical parallel operations. In some cases, the 'renewable power' generated can be sold to the electric utility grid at a premium price, though this is rarely the most cost-effective approach, due to the low export prices most electric utilities offer.

Energy demand-side mechanisms and energy tariff structures are a global trend and should be included in the evaluation of technologies and operational strategies (e.g., process control solutions). Nowadays a gap still exists between energy consumption and costs, since there is no generalized cost model describing current energy tariff structures to evaluate operating costs at WWRFs. The energy market is decentralized, using utility-specific or client-specific accounting functions to calculate energy bills. Within this context, a generalized cost model covering the major energy tariff terms enables planning engineers to: (1) highlight critical situations where peak power demand charges are raising total energy costs; (2) develop strategies to reduce energy consumption on a time-of-use basis and maximize energy production at peak periods; (3) specify the appropriate equipment to reduce overall energy consumption and power demand; (4) identify the critical terms in the energy bill and develop operating strategies to operate and control the plant for their reduction; and (5) find the optimal contracted power capacity structure for a specific plant.

This chapter is adapted from the work of Aymerich *et al.* and Rieger *et al.*, who first demonstrated the connection of tariff structures and process dynamics, and generalized the power structure including the most common billing terms (Aymerich *et al.*, 2015; Rieger *et al.*, 2014).

6.2.2 Energy tariff structures

A large variability of energy tariff structures can be found depending on: (1) the customer category (i.e., residential or industrial, small or large customers); (2) the specific energy pricing structure applied; and (3) the different billing terms involved in the bill. In this section we describe the concepts of energy pricing structures and billing terms behind the most common energy tariff structures for large energy customers based on selected energy contracts obtained from different WRRFs in Europe and North America.

Usually the three main categories of clients can be distinguished among energy suppliers: primary, secondary and sub-transmission clients. Secondary clients are those receiving a service metered and delivered at voltages higher than 2 kV. Secondary clients are mainly residential and small business. WRRF are usually considered primary clients as they are included in the category where energy can be delivered at voltages higher than 2 kV and below 50 kV. Nevertheless, owing to the large energy intensity of many facilities and the need for a reliable network, many large WRRF supplies are delivered in a different category: the sub transmission. Such category is characterised by having special treatment, as the electrical connection is not shared with any other users and is exclusive to the client, making it more reliable and providing direct contact with the energy supplier. This category services voltages up to 66 kV.

6.2.2.1 Energy pricing structures

The energy pricing structure defines how the various charge rates are applied to the different terms of the bill, such as the charges related to the energy usage (expressed for example in € kWh^{-1}), contracting a specific power capacity (expressed as € kV^{-1} or € kW^{-1}) or peak power demand penalties (expressed as € kW^{-1}). Descriptions of the three types of energy pricing structures identified are described below.

6.2.2.2 Flat-rate structure (also called constant or fixed rate)

In a flat-rate structure customers are charged the same amount for the energy they use or peak power they demanded, no matter the time of the day or the quantity that is consumed. This is the simplest structure but rarely applied in energy contracts for large energy customers (e.g., WRRFs).

6.2.2.3 Time-of-use rate structure (TOU)

TOU rate structures are widely applied at utilities across the United States and Europe. In a TOU rate, customers are charged a different price according to the time of day, day of the week and/or season of the year. Figure 6.2 shows a conceptual example of a typical TOU rate structure where different rates are applied depending on the time of day. Normally, in a TOU rate two or three price periods are applied and classified as On-peak or Peak (highest energy price of the day), Mid-peak or Shoulder, and Off-peak (lowest energy price of the day). The On-peak and Mid-peak periods are usually applied during the day (when the highest energy demand occurs), and the Off-peak periods during the night. The mechanism encourages customers to shift their power demand from peak periods (with high prices) to Off-peak periods (with low prices). On the other hand, charge rates applied during the day can vary depending on the season or month. In the majority of the evaluated cases, a winter and summer TOU tariff schedule is defined such as in the US (e.g., Southern California Edison), where different prices and periods are applied in winter and summer, respectively. In other cases, the TOU tariff schedule can change depending on the month, such as in Spain (Royal Decree 1164/2001). TOU rates are of special interest for WRRFs since usually high energy usage and power demand is linked to high load periods, usually coinciding with the highest energy price periods.

6.2.2.4 Tiered rate structure

In a tiered structure (also called step rate or block rate), customers are charged a different price based on the amount of energy used or the maximum peak power demand claimed. Various tariff blocks are defined, where each block is charged at a different price. In this way, when companies have reached the cap of their first block, any additional electricity used is charged at their second block price and so on. Depending on the type of tariff contracted, prices can increase (i.e., tiered rate) or decrease (i.e., inverse tiered rate) for the amount of energy consumed. Examples of tiered rate structures can be seen in the US (e.g., Direct Energy Business), Canada (e.g., Hydro One) and Australia (e.g., Energy Australia).

Figure 6.2 Typical tariff structure with baseline charge, shoulder, and peak (adapted from Rieger *et al.*, 2014): (a) summer season time-of-use energy pricing (from June to September); (b) winter season time-of-use energy pricing (from May to October).

6.2.3 Billing terms: understanding the electrical bill

The electrical bill that customers receive includes several terms, which may vary according to the specific energy tariff structure contracted. The five most common terms are summarized below, as can be seen in Figure 6.3.

6.2.3.1 Fixed charges

The fixed charges (also referred to as customer charges, fixed fee, fixed standing charges, or metering charges) usually cover the costs of access, metering, meter reading, billing and other customer-related operating costs. The fixed charges for

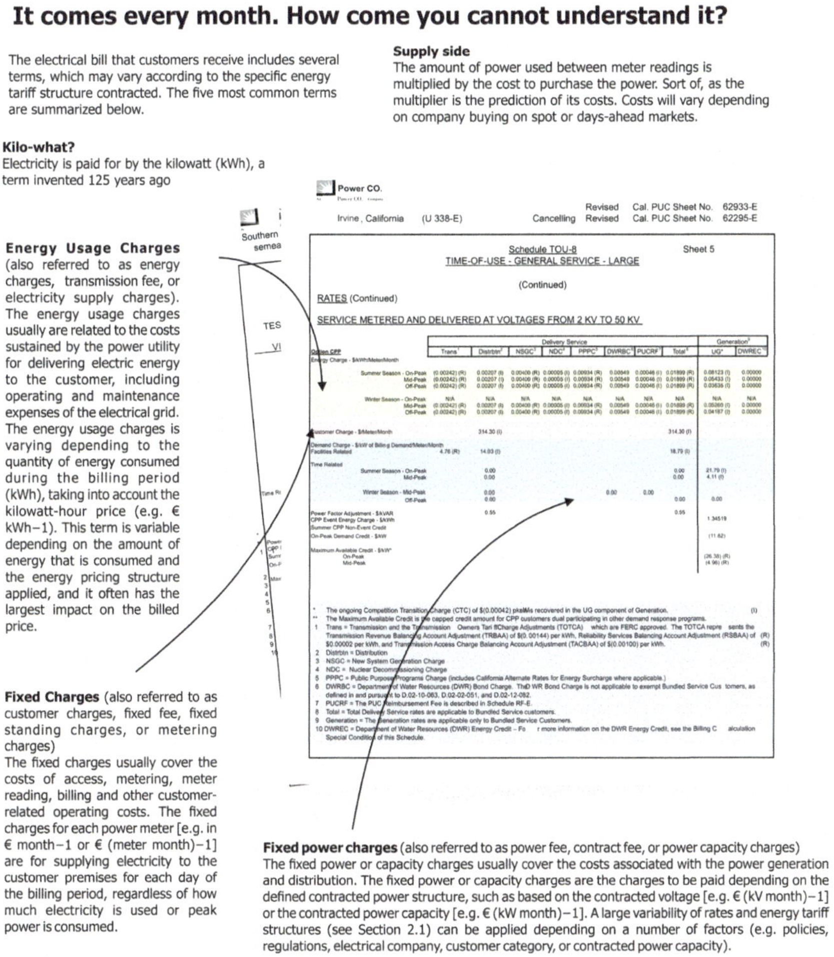

Figure 6.3 One example of energy bill (e.g., US) and related components.

How to calculate your bill

Energy bills have two main components. Often Two separate bills can be received for each of them. The costs relates to price of delivering the required energy, plus the fixed and power charges, and the costs related to produce or generate the energy.

The price will is calculated as follows: **Energy usage charges** amounts to block 1 (delivery) and block 2 (generation) to each kWh serviced, plus **fixed charges** adding block 3 (delivery), plus **fixed power charge** block 4 (delivery) and 5 (generation) for each kV contracted, plus **reactive energy charges** block 6 and plus **other charges** block 7.

(Continued)

RATES (Continued)

SERVICE METERED AND DELIVERED AT VOLTAGES FROM 2 KV TO 50 KV

Delivery | Generation

	Delivery Service								Generation[9]		Block
Option CPP	Trans[1]	Distrbtn[2]	NSGC[3]	NDC[4]	PPPC[5]	DWRBC[6]	PUCRF[7]	Total[8]	UG*	DWREC[10]	
Energy Charge - $/kWh/Meter/Month											
Summer Season - On-Peak	(0.00242) (R)	0.00207 (I)	0.00400 (R)	0.00005 (I)	0.00934 (R)	0.00549	0.00046 (I)	0.01899 (R)	0.08123 (I)	0.00000	
Mid-Peak	(0.00242) (R)	0.00207 (I)	0.00400 (R)	0.00005 (I)	0.00934 (R)	0.00549	0.00046 (I)	0.01899 (R)	0.05433 (I)	0.00000	
Off-Peak	(0.00242) (R)	0.00207 (I)	0.00400 (R)	0.00005 (I)	0.00934 (R)	0.00549	0.00046 (I)	0.0[illegible] (R)	0.0[illegible] (I)	0.00000	
Winter Season - On-Peak	N/A	N/A	N/A	N/A	N/A	N/A	N/A	1	2	N/A	1, 2
Mid-Peak	(0.00242) (R)	0.00207 (I)	0.00400 (R)	0.00005 (I)	0.00934 (R)	0.00549	0.00046 (I)	[illegible] (R)	0.[illegible]	0.00000	
Off-Peak	(0.00242) (R)	0.00207 (I)	0.00400 (R)	0.00005 (I)	0.00934 (R)	0.00549	0.00046 (I)	0.01899 (R)	0.041[illegible] (I)	0.00000	
Customer Charge - $/Meter/Month		314.30 (I)						314.30 (I)			3
Demand Charge - $/kW of Billing Demand/Meter/Month											
Facilities Related	4.76 (R)	14.03 (I)						18.79 (I)			4
Time Related											
Summer Season - On-Peak		0.00						0.00	21.79 (I)		5
Mid-Peak		0.00						0.00	4.11 (I)		
Winter Season - Mid-Peak		0.00				0.00	0.00				
Off-Peak		0.00						0.00	0.00		
Power Factor Adjustment - $/kVAR		0.55						0.55			
CPP Event Energy Charge - $/kWh									1.34519		6
Summer CPP Non-Event Credit											
On-Peak Demand Credit - $/kW									(11.82)		
Maximum Available Credit - $/kW*											
On-Peak									(26.38) (R)		7
Mid-Peak									(4.96) (R)		

Peak Power Demand (also referred to as demand, distribution demand, penalty, or overuse charges)
Peak power demand charges are common demand side management mechanisms used to cover the extra costs for excessive power consumption within a specified short period of time. The peak power demand charges are usually based on the maximum peak power demand (kW) measured in any time interval (e.g. 15min, 30min, or 60min),

Reactive Energy Charges (also referred to as reactive power factor or reactive power fee)
The reactive energy charges cover the costs for the energy or power dissipated by inductive electrical equipment, measured as reactive energy (kVArh) or reactive power (kVAr). The reactive energy charges are referred to the price that has to be paid when there is an excess consumption of reactive energy or power. In other words, the reactive energy charges are the charges for the inefficiency at the customer's site. The level of inefficiency is usually expressed as a percentage and is called power factor (ratio between active power and apparent power).

Time-of-use rate structure (TOU)
TOU rate structures are widely applied at utilities across the United States and Europe. In a TOU rate, customers are charged a different price according to the time of day, day of the week and/or season of the year.

Figure 6.3 *(Continued)*

each power meter [e.g., in € month^{-1} or € (meter month)$^{-1}$] are for supplying electricity to the customer premises for each day of the billing period, regardless of how much electricity is used or peak power is consumed.

6.2.3.2 Fixed power charges

The fixed power or capacity charges (also referred to as power fee, contract fee, or power capacity charges) usually cover the costs associated with the power generation and distribution. The fixed power or capacity charges are the charges to be paid depending on the defined contracted power structure, such as based on the contracted voltage [e.g., € $(kV\ month)^{-1}$] or the contracted power capacity [e.g., € $(kW\ month)^{-1}$]. A large variability of rates and energy tariff can be applied depending on a number of factors (e.g., policies, regulations, electrical company, customer category, or contracted power capacity).

6.2.3.3 Energy usage charges

The energy usage charges (also referred to as energy charges, consumption charges, transmission fee, or electricity supply charges) are usually related to the costs sustained by the power utility for delivering electric energy to the customer, including operating and maintenance expenses of the electrical grid. The energy usage charges vary depending on the quantity of energy consumed during the billing period (kWh), taking into account the kilowatt-hour price (e.g. € kWh^{-1}). This term is variable depending on the amount of energy that is consumed and the energy pricing structure applied (see Sec. 6.2.1), and it often has the largest impact on the billed price.

6.2.3.4 Peak power demand charges

Peak power demand charges (also referred to as demand, distribution demand, penalty, or overuse charges) are common demand-side management mechanisms used to cover the extra costs for excessive power consumption within a specified short period of time. The peak power demand charges are usually based on the maximum peak power demand (kW) measured in any time interval (e.g., 15 min, 30 min, or 60 min), in most cases during a monthly billing period or during the previous 11 months, such as in the United States (e.g., Dominion Virginia Power VEPGA). In the majority of cases such as in the United States (e.g., Southern California Edison) or in Sweden (e.g., Vattenfall), the peak power demand charges are determined based on the maximum peak power demand measured in a billing period. In other cases, such as in Spain (Royal Decree 1164/2001), in the United States (e.g., Dominion Virginia Power VEPGA), in Sweden (e.g., E. ON Energy Company, Tekniska Verken) or Canada (e.g., Hydro Quebec), the peak power demand charges are adjusted based on the difference between the maximum peak power consumed and the contracted power capacity.

6.2.3.5 Reactive energy charges

The reactive energy charges (also referred to as reactive power factor or reactive power fee) cover the costs for the energy or power dissipated by inductive electrical equipment, measured as reactive energy (kVArh) or reactive power (kVAr). The reactive energy charges are referred to the price that has to be paid when there is an excess consumption of reactive energy or power. In other words, the reactive energy charges are the charges for the inefficiency at the customer's site. The level of inefficiency is usually expressed as a percentage and is called power factor (ratio between active power and apparent power). In cases such as in Spain (Royal Decree 1164/2001), the level of inefficiency is expressed as a function of the cos(φ) value, where φ is the angle of difference (in degrees) between the active power and apparent power, which is the quantification of the departure between 1.0 (ideal condition where only non-reactive power is drawn or the electrical system is fully re-phased) and the actual customer condition (<1.0). These charges are site-specific and respond to the properties and status of the electrical equipment. A way to reduce or eliminate reactive energy charges can be by installing, e.g., capacitors or replacing existing equipment (e.g., motors, transformers, or other energy consumers) with more energy-efficient equipment.

6.2.3.6 Taxes

Taxes are site-specific and can include: customer taxes, energy commission taxes, regulatory taxes, delivery taxes, or green energy taxes.

6.2.4 Benefits of understanding energy tariff structures

A recent study showed the importance of incorporating realistic energy cost models based on current and local-based energy tariff structures into WRRFs (Aymerich *et al.*, 2015). The study modelled a typical WRRF serving a population of 115,000 at an average flow of 18,000 $m^3\ d^{-1}$ as a case study. The WRRF was modelled in SIMBA# (ifak e.V., Germany) and using the Benchmark Simulation Model (BSM) principles (Gernaey *et al.*, 2014). For the case evaluated, monthly cost differences of 7–30% were observed compared to the proposed realistic energy cost model with a DO PI control strategy. It was demonstrated that using average energy prices and neglecting energy tariff structures may lead to biased conclusions when selecting operating strategies (e.g., control solutions) or comparing technologies or equipment. It is important to note that selecting the optimal power contracted is also a key issue, since different operating strategies result in different optimal contracted power, hence proper and site-specific energy costs models are required.

6.3 DYNAMICS OF AERATION ENERGY

WRRFs are subject to continuous disturbances caused by influent variations, which exhibit diurnal patterns as well as stochastic changes due to storm events. Daily patterns are the translation of population activity dynamics during the course of the day, resulting in disturbances in both flow rate and concentration, inevitably changing the DO requirements. Thus, the combination of urban, commercial and industrial dynamics subjects WRRFs to significant changes in wastewater characteristics, influencing the overall energy-associated costs.

In general, the variations in COD, BOD, TKN and TSS follow the pattern of flow-rate variations (Metcalf & Eddy, 2014). The time and amplitude of the peaks will depend on the type of constituent under study, size and characteristics of the community, and/or on whether is weekday or weekend. Often, differences of a factor of ten between the lowest and the highest load are reported. During peak time, wastewater treatment requires higher rates of aeration to maintain the DO set points and to comply with effluent water quality standards. Satisfying the requirements of a changing DO implies a need for aeration blowers to adapt by adjusting the amount of air supplied into the wastewater. Thus, as aeration blowers are the core of the energy demand, the corresponding equivalent associated energy, costs, and greenhouse gas emissions will also be subjected to the changing characteristics of the influent. Even changes in the carbonaceous fractionation of the wastewater constituents (i.e., readily and slowly biodegradable organic matter, etc.) will have an impact on the overall energy requirements (Mannina *et al.*, 2016).

The energy requirements during circadian cycles will be further exacerbated due to various factors. Recent studies demonstrated that daily variations in COD, TKN, influent flow, and αSOTE, together with energy tariff structures and current energy production methods exacerbated air requirements and GHG emissions for WRRFs sited in areas where peak plant load corresponds to peak power rates. The compounding effect of peak flow, peak load, and dynamics of the aeration system performance needs to be considered. This amplification is illustrated in Figure 6.5 and is described in the following sections.

6.3.1 Circadian amplification of air requirements

The oxygen transfer efficiency fluctuates through daily cycles because of the effects of wastewater contaminants (Jiang *et al.*, 2017; Leu *et al.*, 2009)

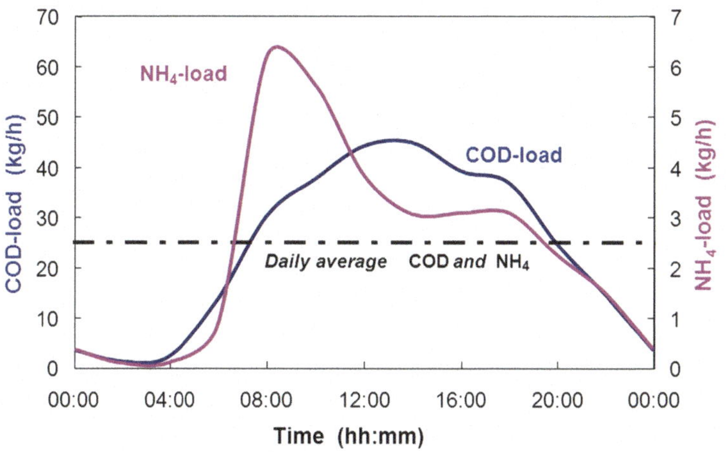

Figure 6.4 Typical hourly load variation (i.e., COD and ammonium) in domestic wastewater resource facility. Adapted from Urban Water Systems by Peter Krebs, Technische Universität Dresden, Department of Hydro Science, Institute for Urban Water Management.

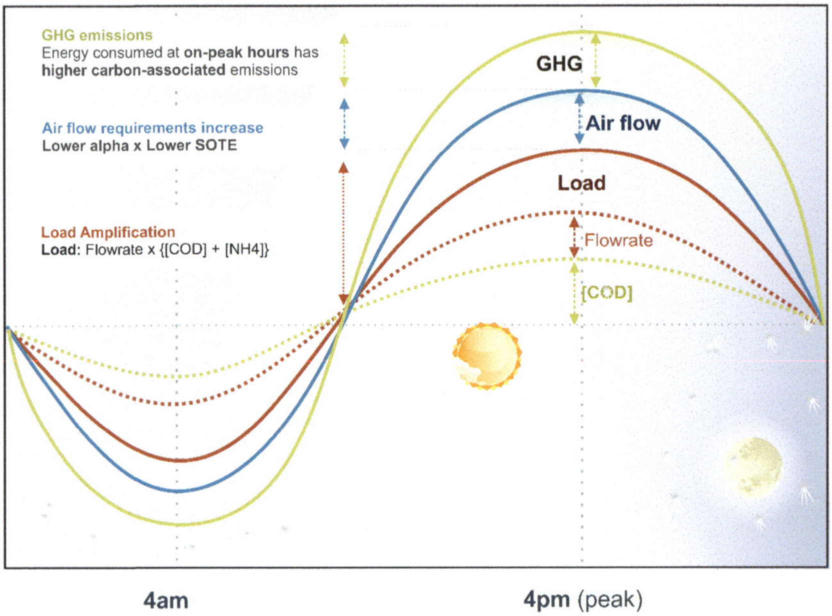

Figure 6.5 Illustration of the circadian amplification in WRRFs where peak load and peak power rates compound.

(Fig. 6.4). High concentrations of contaminants such as surfactants and biodegradable COD (bCOD) decrease the α factor (see Sec. 2.1.3), depressing oxygen transfer efficiency (Leu *et al.*, 2009; Rosso & Stenstrom, 2006b). Under normal conditions, high constituent concentrations also coincide with the highest hydraulic loadings, thus driving aeration efficiency to its lowest value when the oxygen demand (energy) is the highest.

Moreover, when there is high oxygen demand due to high organic loadings, the aeration system will have to adapt and increase the airflow to satisfy DO requirements. As airflow increases, two physical effects further reduce the oxygen transfer from gas to liquid phase (Rosso *et al.*, 2005): the combined net effect of (1) formation of larger bubbles, lowering the surface to volume ratio; and (2) increase in bubble rise velocity, reducing the contact time for oxygen mass transfer.

6.3.2 Circadian amplification of energy-associated GHG emissions

The reduction of gas transfer efficiency with increased pollutants concentration is certainly unfortunate. Nevertheless, in addition to the increase in energy requirements, the situation still can worsen if we consider the indirect energy-associated GHG emissions. The GHG or carbon intensity of the unit power generation (g_{CO2}/kWh) varies in the diurnal period independently from the water characteristics. Indirect carbon emissions also face variations among the portfolio of energy sources employed by power supply agencies in different hours of a diurnal period. During the night, the carbon intensity of unit power generation is lower, due to a higher contribution of clean-source energies; while during the daytime, when activity peaks are found, fossil fuel source power has a bigger contribution in fulfilling the augmented power demand. Therefore, the period of greatest aeration rate often tends to overlap with the highest energy-associated emissions. Different studies confirmed and quantified that the amplification of energy consumption during peak hours meant an increase in energy-associated carbon emissions (Emami *et al.*, 2018; Sobhani, 2011).

In Figures 6.6 and 6.7, a diurnal emission analysis is reported. These results were obtained using the estimated carbon emission intensity at different times during the day. Indirect GHG emissions from production of electricity will vary spatially and temporally across the United States, depending on the energy sources (Graff Zivin *et al.*, 2014).

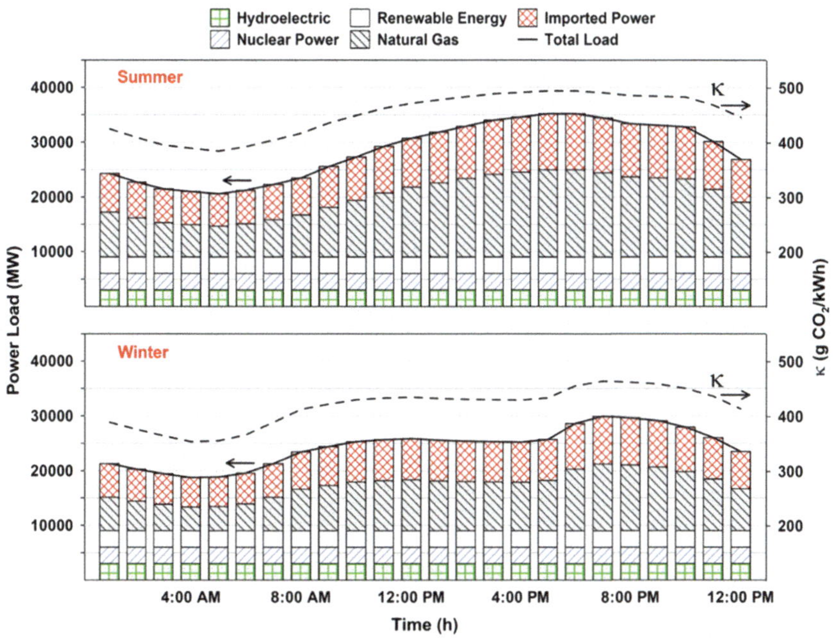

Figure 6.6 Diurnal variation of power load (MW) and its associated carbon emission (Kg_{CO2}/kWh) for summer (top) and winter (bottom) emissions for Southern California in 2010.

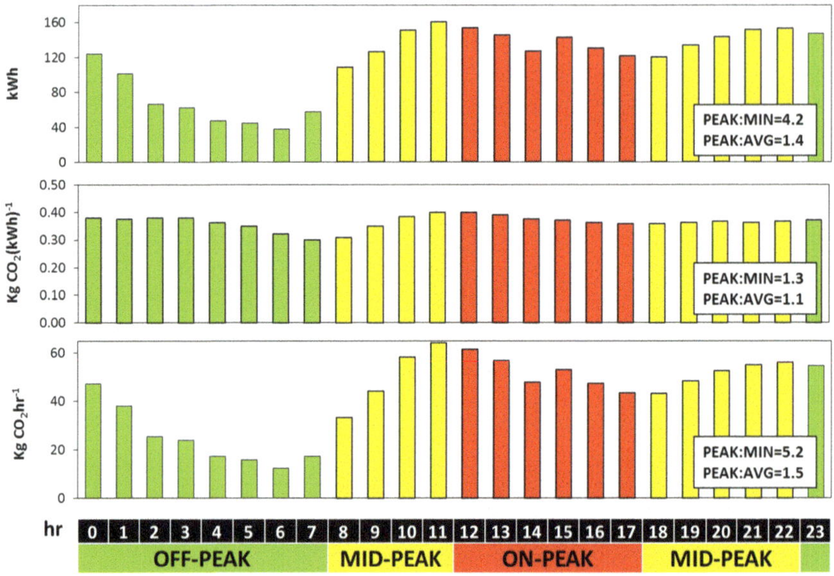

Figure 6.7 Circadian amplification of TOU power and CO_2 emissions. Adapted from Emami *et al.* (2018).

6.4 ENERGY CONSEQUENCES OF INEFFICIENT OR FOULED EQUIPMENT

6.4.1 From efficient to inefficient fine-pore diffusers

Continuous research on developing more efficient aeration systems has found fine-pore diffusers as a promising avenue to reduce the energy required due to their higher aeration efficiency. Since the 1970s, fine-pore diffusers have become the most common diffusers employed in aeration systems (Aberley *et al.*, 1974), thus, presenting a competitive advantage in efficiency in comparison to coarse-bubble systems (Henze *et al.*, 2008; Kaliman *et al.*, 2008; Rieth *et al.*, 1990). Fine-pore diffusers are characterized by large surface areas to distribute a myriad of small bubbles through the water column (i.e., with diameter less than 5 mm). Due to their large surface area and the relatively low shear from the bubbles released, these diffusers offer an opportunity for microbial adhesion and growth, providing a unique habitat for biofilm formation characterized by high availability of oxygen, a wide range of substrate, and a surface (often biocompatible) for adhesion. Therefore, once installed all diffusers progressively and inexorably lose their initial performance (Gillot *et al.*, 2005; Hertle, 2015; Kim & Boyle, 1993; Wagner & Pöpel, 1998). Biological fouling has the double-negative effect of diminishing the oxygen transfer efficiency as bubble coalescence results in fewer but larger bubbles (i.e., higher velocity from the bottom to the top of the bioreactor, lower contact time, lower specific area) and promoting additional pressure drop through the diffuser (Frey & Thonhauser, 2004; Henze *et al.*, 2008; Kaliman *et al.*, 2008; Rosso & Stenstrom, 2006b; Wagner *et al.*, 2007), which together compound an increase in aeration energy intensity to meet the same oxygen requirements. This results in an increase of the energy required for supplying the same amount of oxygen over time, which eventually reduces up to 50% of its original efficiency (Garrido-Baserba *et al.*, 2017; Rosso *et al.*, 2005). Thus, since aeration is identified as the prime contributor to energy use in WRRFs, the escalation of energy wastage and the deterioration of economic sustainability at reclamation plants will largely depend on fouling (Henze *et al.*, 2008; Kim & Boyle, 1993; Redmon *et al.*, 1983; Rosso & Stenstrom, 2007).

6.4.2 Understanding fouling

Diffusers offer a unique habitat for biofilm formation: high availability of oxygen, a wide range of substrates, and a microbial-friendly surface for adhesion. Convective flows transport particulates (including bacterial cells) from the mixed liquor to the diffuser surface. Both the substrate from the incoming wastewater and the soluble microbiological products (SMP) produced by the planktonic biomass in the aeration tank are readily available for the biofilm, serving as a substrate for the sessile microbial community. The exact microbial composition and its extent will vary from one treatment plant to another, as biofilm formation is affected by operating-process variables (i.e., availability of soluble substrate, dissolved oxygen level, iron concentration in the mixed liquor, organic loading rate; Kim & Boyle, 1993; Rieth *et al.*, 1990).

The biofouling layer is a composite of bacterial flocs, protozoa, soluble microbial products and extracellular polymeric substances (EPS), particulate and soluble substrate and inorganic scales (Figure 6.8). The works of Chao *et al.* (2014) suggested that WWTP biofilms could be composed of different types of EPS (e.g., firm and loose), which would result in specific microbial consortium compositions with different genera and phyla, with each providing a different set of characteristics or functions to the consortium. The exact role of each type of biofilm and the dynamics among phyla, genera, and different types of EPS are still poorly understood. Even less knowledge is available regarding the capacity of the microbial biofilm to deplete the material compounds at the diffuser surface. As biofilm density increases, organisms at the surface may encounter limited nutrients, thus triggering the mining of the fine-pore surface substrate such as softener (which primarily consists of organic material). Wagner and von Hössle (2004) have shown how softener content in used EPDM (ethylene propylene diene monomer) membranes can be degraded by as much as 74% when compared to new EPDM diffusers. Nevertheless, site-specific conditions hinder the extrapolation of these findings, as in several cases WWTPs can operate EPDM-type diffusers for almost 15 years (Kaliman *et al.*, 2008; Rosso & Shaw,

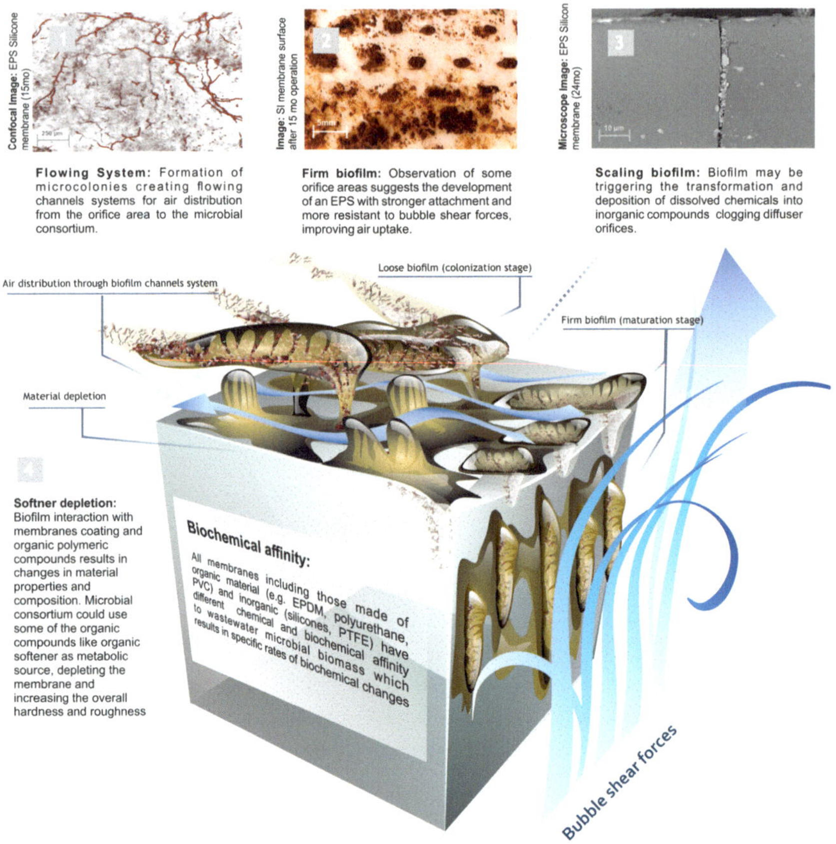

Figure 6.8 Illustration of biofilm interaction with mechanical and biomechanical fine-pore diffuser properties at the orifice surface of fine-pore diffusers, with micrographs showing details of the EPS structure, biofilm attachment and scaling phenomena. Adapted from Garrido-Baserba *et al.* (2017).

2015; Wagner & von Hössle, 2004). Commercially available fine-pore diffusers may be coated or polycoated to provide or enhance fouling resistance, but there is still a general lack of independent verification of the extent of fouling resistance. It is suspected that bacterial hydrophobicity and the biocompatibility of membrane-like surface of fine-pore diffusers could facilitate biofilm formation, including bacterial initial attachment as well as the subsequent development of the bacterial colony (Palmer *et al.*, 2007).

6.4.2.1 Linking aeration efficiency and molecular biology

Although the importance of microbial biofilms regarding energy efficiency in WRRFs, which influence approximately 45–75% of the total energy costs, their

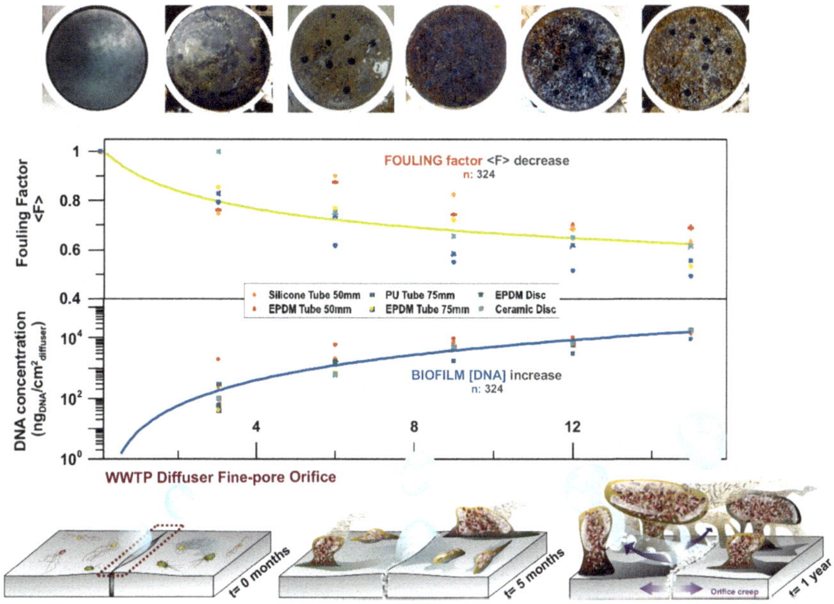

Figure 6.9 Correlation between total biofilm DNA concentration and the decrease in diffuser performance (expressed by the fouling factor *F*); Adapted from Garrido-Baserba *et al.* (2017).

dynamics and composition have been largely ignored. Although the connection between biofilm growth and fouling has been implicit in discussions of diffuser fouling for many years, recent studies provided directly measured evidence of the connection between biofouling and reduced diffuser efficiency (Garrido-Baserba *et al.*, 2017; Rosso & Shaw, 2015). Those studies quantified how the performance of aeration diffusers in wastewater (i.e., αF, and αSOTE) was consistently eroded by biofilm growth on the diffuser surface, providing a mechanistic correlation between the decreased oxygen transfer efficiency and biofilm growth (Figs. 6.9 and 6.11).

A recent study systematically characterized and compared the biofilm dynamics and aeration performance of different fine-pore diffusers in activated sludge. Six diffusers were selected and operated continuously for 15 months in a pilot reactor. Molecular techniques were used to quantify and examine the microbial structure in the biofilms retrieved from the diffuser surface at different points in time. The study confirmed how the aeration efficiency parameters steadily decreased during the time in operation (Figure 6.10). The values of the fouling factor decreased by approximately 25%, showing a $<F>$ average of 0.75 ± 0.05 for all diffusers for the studied period.

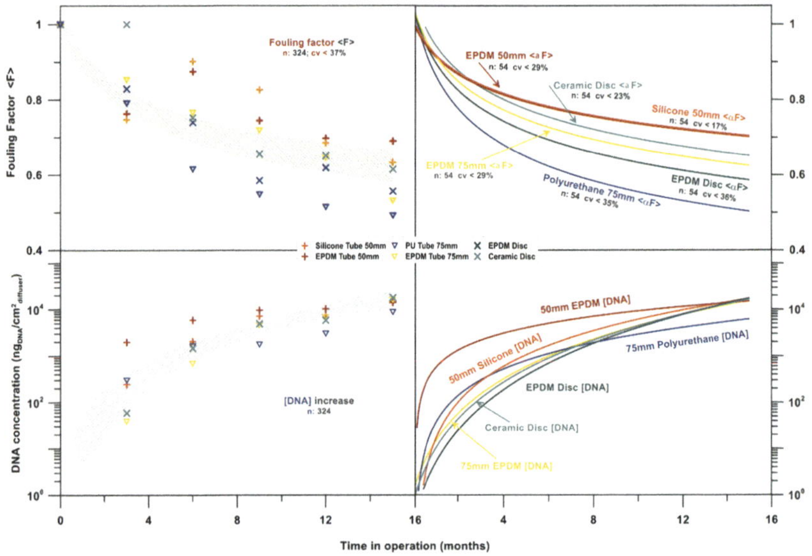

Figure 6.10 Diffuser fouling and biofilm increase (expressed as DNA concentration) for six fine-pore diffusers over the project period. The fouling values are flow-weighted averages for each diffuser at each time of measurement. Also, for each diffuser, each DNA point represents the average DNA concentration value obtained from nine replicate samples for each testing period.

The progressive decrease in aeration efficiency during operation tends towards a stabilization, as the development of biofilms starts to plateau. However, due to the time limitations of the study a plateau had not been reached by the end of the study. Nevertheless, by extrapolating to an asymptote, the rate of fouling decline seemed to stabilize at <*F*> ~0.5 after 24–30 months. Similarly, the biofilm growth rate also seems to stabilize at $t = 24$ mo. Such results are in accordance with Rosso and Stenstrom (2006b), who analysed datasets from 94 field measurements and showed a clear pattern of performance decline with time in operation. The study could confirm that for those WRRFs using fine-pore diffusers, efficiency tends to decline rapidly during the first 24 months of operation and that any diffusers remaining in operation indefinitely may end up performing at less than 50% of original efficiency. Although from this observation a conclusion in absolute terms should not be drawn, because the <*F*> is a measure of relative efficiency (for used vs. new diffusers), and hence it is a measure of consistency and not necessarily of absolute efficiency. In fact, diffusers that in general tend to operate at high air flux, thereby releasing larger bubbles, exhibit stable and higher <*F*> values, indicating that their efficiency (expressed as αFSOTE) maintains a lower but consistent value over time.

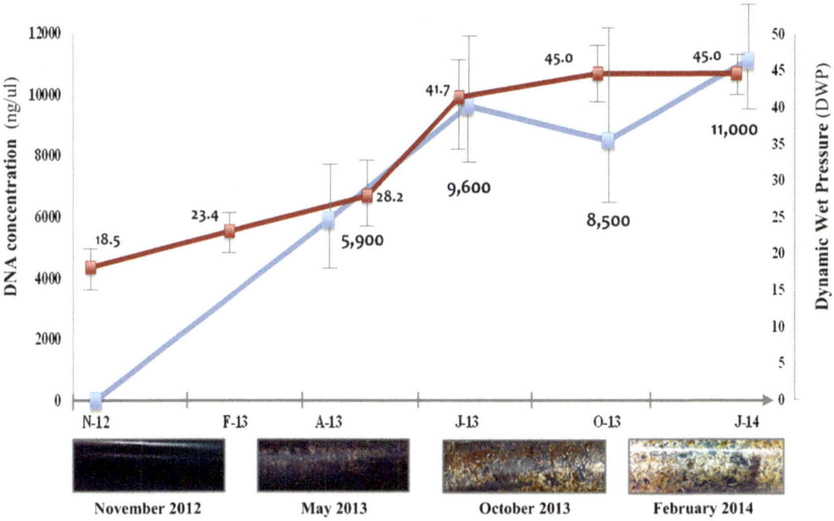

Figure 6.11 Images of the EPDM tube membrane with corresponding average DNA mass per unit volume and dynamic wet pressure. Results are from averaging nine data points from the top, side and bottom sections of the tube membranes (adapted from Rosso & Shaw, 2015).

By plotting the top and bottom graphs in Figure 6.10 against each other, Figure 6.12 can be produced. It can be seen that the initial performance in process water (i.e., αSOTE) is eroded by biofilm growth, quantified by the DNA concentration in the biofilm colonizing the diffuser surface. The average coefficient of variance regarding the correlation between biofilm DNA and the decrease in diffuser performance (expressed by αF) for the six diffusers was 14 $\pm$ 2% and it never exceeded 36%. The coefficient of variance in Si and ceramic diffusers never exceeds 17%, whereas the EPDM disc (with an average CV of 20 $\pm$ 4) exhibited the highest variance, 36%. Therefore, Figure 6.12 indicates the decrease in $<F>$ is directly correlated to the increase in biofilm growth, capturing and quantifying the parallel decrease of $<F>$ together with the increase on DNA concentration. On average an increase of 2-log in DNA concentration diminished $<F>$ by 20%, whereas a 3-log increase reduced the efficiency parameter by nearly 30% and approximately 40–45% when reaching what seems to be a plateau around the 4-log concentration. Diffuser geometry showed slight influence in the correlation.

6.4.3 Influence of the organic load in biofouling

The formation of biofilms in aerated processes is influenced by operating process variables (i.e., solid retention time, availability of soluble substrate, dissolved

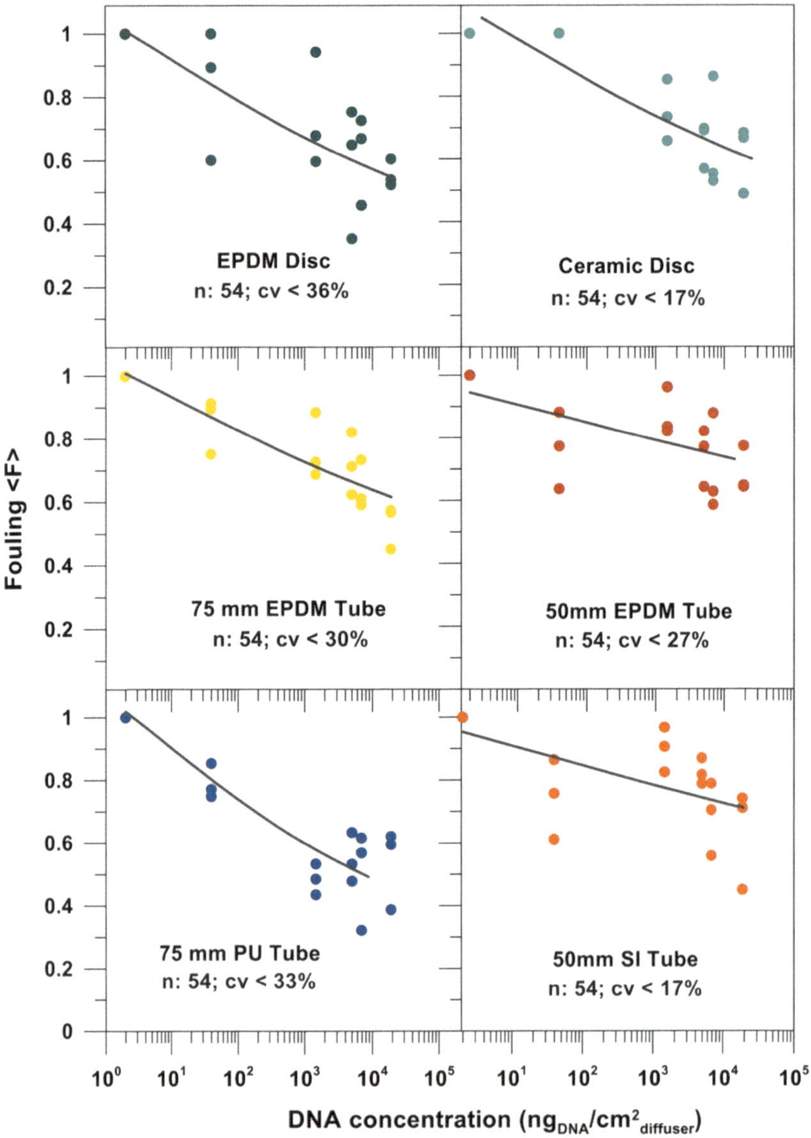

Figure 6.12 Correlation between total biofilm DNA concentration and the decrease in diffuser performance (expressed by αF) for the six membranes. For each plot the coefficient of variation never exceeded the value reported.

oxygen level, organic loading rate) (Kim & Boyle, 1993; Rieth *et al.*, 1990). However, there is no clear evidence yet regarding how these operational conditions together with diffuser membrane properties (i.e., material type, surface roughness, etc.) are correlated with the fouling factor and the dynamic nature of biofilm microbial communities. In order to provide a better understanding of the influence of operational conditions in fine-pore type diffusers, two replicates of different aeration systems (same manufacturer, material, and model) were installed and tested in two different operational conditions (higher and lower rate processes). Two types of loading rate regimes were identified as the main factors influencing αF values after a statistical analysis. High rate process conditions decreased aeration efficiency by more than 20% in comparison to conventional treatment process (Figures 6.13 and 6.14) (Garrido-Baserba *et al.* 2017).

The main hypothesis was that higher process rate conditions enhance the production of EPS, which results in a more aggressive clogging of the diffuser pores. However, this theory was discarded after EPS samples taken along the time in operation failed to support it. Therefore, the observed decrease in efficiency parameters seems to be related to higher biocompatibility between microbial consortium and diffuser surface. The higher-rate conditions promote a

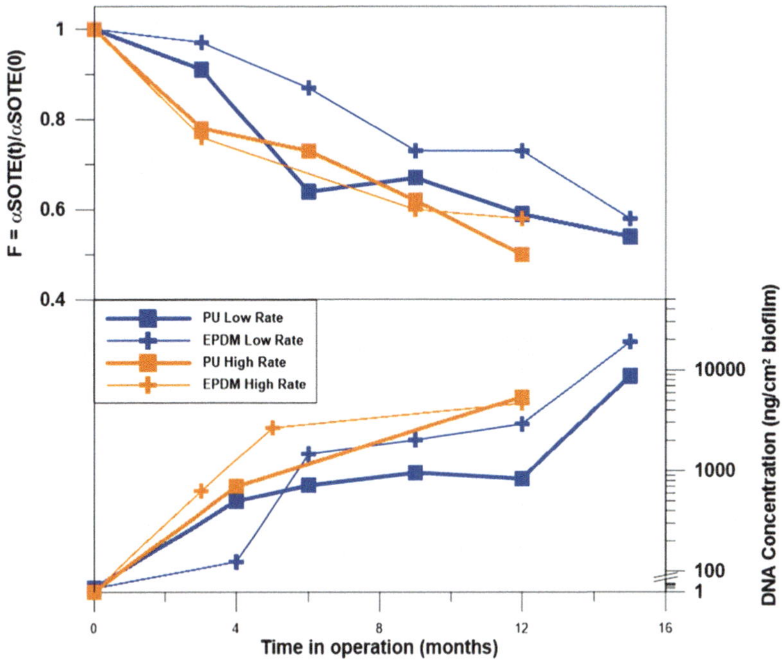

Figure 6.13 Effects of different influent rates on aeration performance for two diffuser membrane materials (square symbol – PU = polyurethane; cross symbol – EPDM) at midrange air flux. Note that the DNA concentration, a quantification of the biofilm mass, is in log-scale.

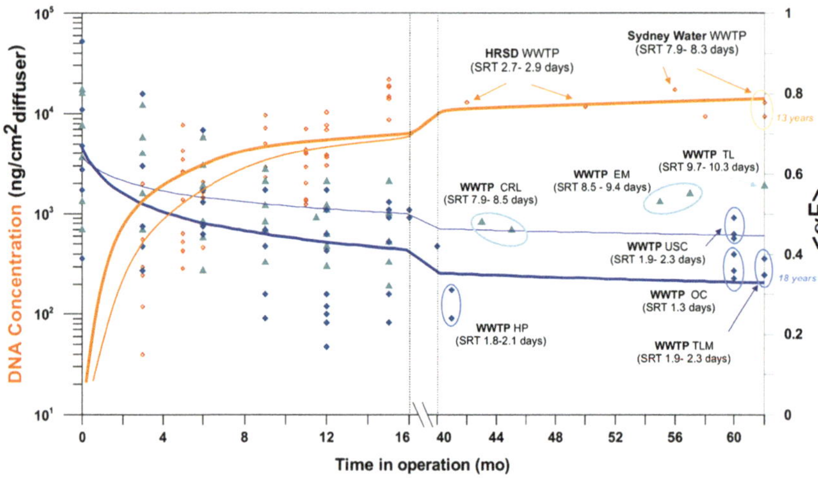

Figure 6.14 Diffuser biofilm increase (expressed as DNA concentration) over the project period (15 mo) for the five fine-pore diffusers at higher-rate conditions (thicker orange) and lower-rate conditions (lighter orange), together with the flow-weighted fouling values at higher-rate conditions (thicker blue) and lower-rate conditions (lighter blue).

highly specialized biofilm, which rapidly changes material properties and can accelerate diffuser aging.

6.4.4 Breaking down the aggregate fouling factor

Although there is an observed correlation between biofilm growth (e.g., expressed as DNA concentration) and decrease of efficiency parameters (i.e., fouling factor), it cannot be considered that biofilm itself is the unique factor behind such a decline. Other phenomena like inorganic scaling or material ageing also contribute to the overall decrease in aeration performance of fine pore diffusers. The concept of fouling is an aggregate parameter of three main contributors: organic (related to microbial growth), inorganic (related to the formation of clogging inorganic compounds, or scaling of carbonates, sulphates, silica, etc.), and aging (related to the progressive material deterioration over time, which may be triggered by intense microbial attachment and substrate depletion):

$$F = F_{\text{org}} \cdot F_{\text{inorg}} \cdot F_{\text{ageing}} \tag{1}$$

The impact of fouling on the progressive deterioration of aeration diffusers properties was evaluated for the first time using molecular tools and state-of-the-art techniques. EDX (Energy Dispersive X-Ray) compositional analysis of the most common diffusers provided valuable information that helped

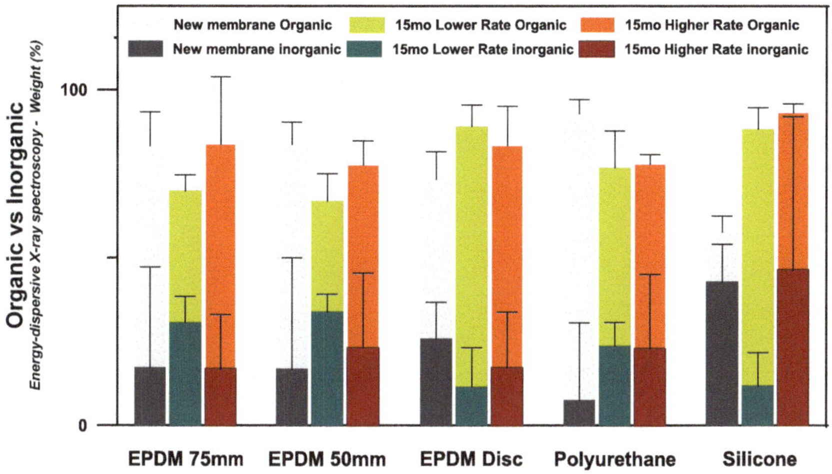

Figure 6.15 Energy-dispersive X-ray spectroscopy results on the compositional analysis for five diffusers, $t = 0$ months (grey/black bars) and $t = 15$ months (green and orange-like bars) in operation.

to break down the specific weight of each of the factors contributing to the fouling (Garrido-Baserba *et al.*, 2018). The results are based on the analysis of a set of diffusers that were in operation for almost 2 years. The EDX analyses (Figure 6.15) showed a biofilm or foulant composition close to 85–90% organic matter (biofouling) and a theoretical maximum of 5–10% inorganic matter (scaling).

6.5 THE LINK BETWEEN PROCESS BIOLOGY AND POWER BILL

6.5.1 Relationship between energy requirements and microbial growth

The final step in linking bacterial fouling and energy wastage by aeration systems can be achieved by relating the biological results with variations in aeration efficiency, and using the latter to calculate the variations in energy requirements to supply oxygen to the process. In order to do so, brake horsepower can be calculated with an adiabatic formula (Metcalf & Eddy, 2014) for blowers to supply air with a time-dependent aeration efficiency and DWP, for all diffusers. Figures 6.16 to 6.17 show the results for the scenario where the oxygen transferred to the water by each diffuser (i.e., OTR/diffuser) is targeted to remain constant, so the diffuser has to release more air to compensate for a decreased oxygen transfer efficiency. The combined effect of decreased SOTE and increased DWP resulted in the ceramic (Figure 6.17), EPDM and polyurethane

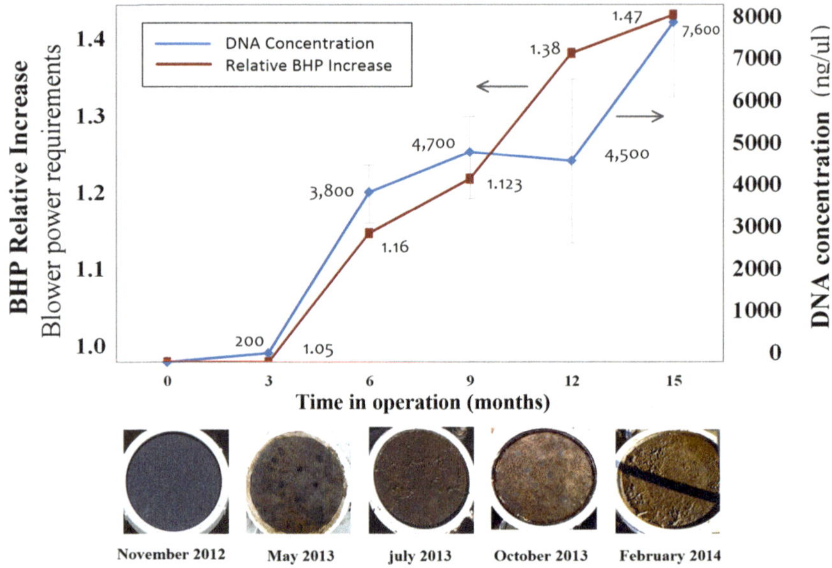

Figure 6.16 Relative increase in power requirements BHP for 9-inch ceramic disc from November 2012 to January 2014.

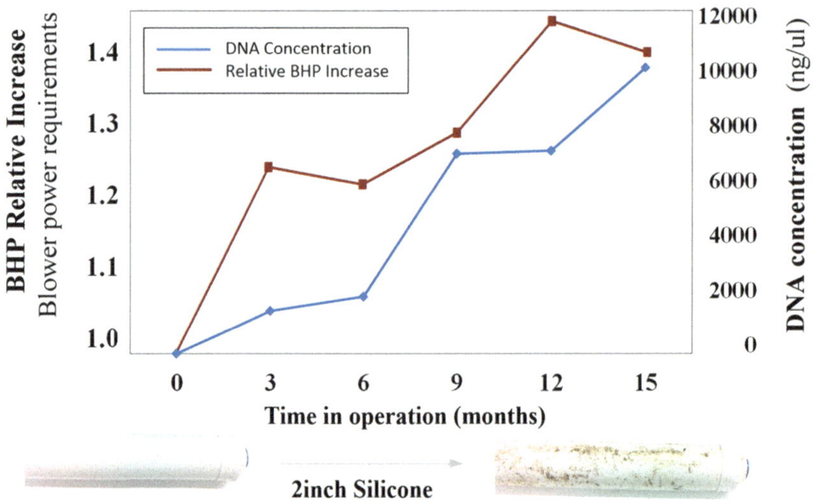

Figure 6.17 Relative increase in power requirements BHP for 2-inch silicone tube from November 2012 to January 2014.

cases requiring over 50% more energy after 15 months than when initially installed. The silicone tube showed a slightly lower increase of just under 30% (Figure 6.17).

6.5.2 Translating the biofouling to energy costs

A direct correlation between biofilm growth (expressed as DNA concentration) and energy requirements was found (Garrido-Baserba *et al.*, 2018). The correlation put in evidence the mechanistic relation between energy demand (and concurrent associated energy costs) and fouling (i.e., the combination of biofouling, scaling, and material ageing). The average coefficient for such correlation was surprisingly significant, with a coefficient of determination of $R^2 = 0.74 \pm 0.07$ (Figure 6.18). The capacity to quantify the decrease in aeration parameters by using biological indicators like EPS, biofilm thickness or amount of DNA could be used to carry out, or at least improve, energy predictions of aeration diffusers.

A recent study used such correlation to develop customized energy projections based on the specific biofilms developed in each type of membrane. Once the specific model equation for each type of membrane was created, the effect of process conditions was incorporated. Process conditions were found to highly influence the rate at which aeration diffusers efficiency deteriorates. Taking into account such influence enabled a simulation of future associated costs and the creation of uncertainty bands in the energy projections. Conservatively, the projections can be applied to those membranes installed in the lower-rate process to provide realistic energy predictions on decreased fouling and, particularly, to determine the cost escalation over time, together with the potential cost savings that cleaning strategies could have on each specific diffuser type.

Figure 6.19 presents conservative energy projections for diffuser operation. The initial 15 months correspond to real data, whereas from 15 to 32 months we present extrapolated values. The uncertainty band was calculated using the observed variations in both (1) process variables (i.e., influent characteristics, expected changes in process conditions, effluent target); and (2) equipment variables (i.e., diffuser type and number, DWP, oxygen transfer efficiency). Conservatively, three intervals of confidence based on these variables were applied: 10% ($t = 15$–21 mo); 12.5% ($t = 21$–26); 15% ($t = 27$–32).

As an example, Figure 6.19 shows the results regarding the projected performance of aeration diffusers (aeration costs increase along with time in operation), and also the overall potential savings that could be theoretically achieved for each type of diffuser. In the case of EPDM discs, it can be seen that it has a lower initial energy intensity than the ceramic disc (103 kWh/1000 m^3 and 134 kWh/1000 m^3, respectively). However, EPDM disc experiences a higher energy consumption increase at $t = 15$ mo (27% vs. 22%, respectively).

The developed methodology was applied to different types of diffusers using the collected data of other types of diffusers and their corresponding biological indicators (i.e., EPS, bacterial concentration, DNA, etc.). Once all models were

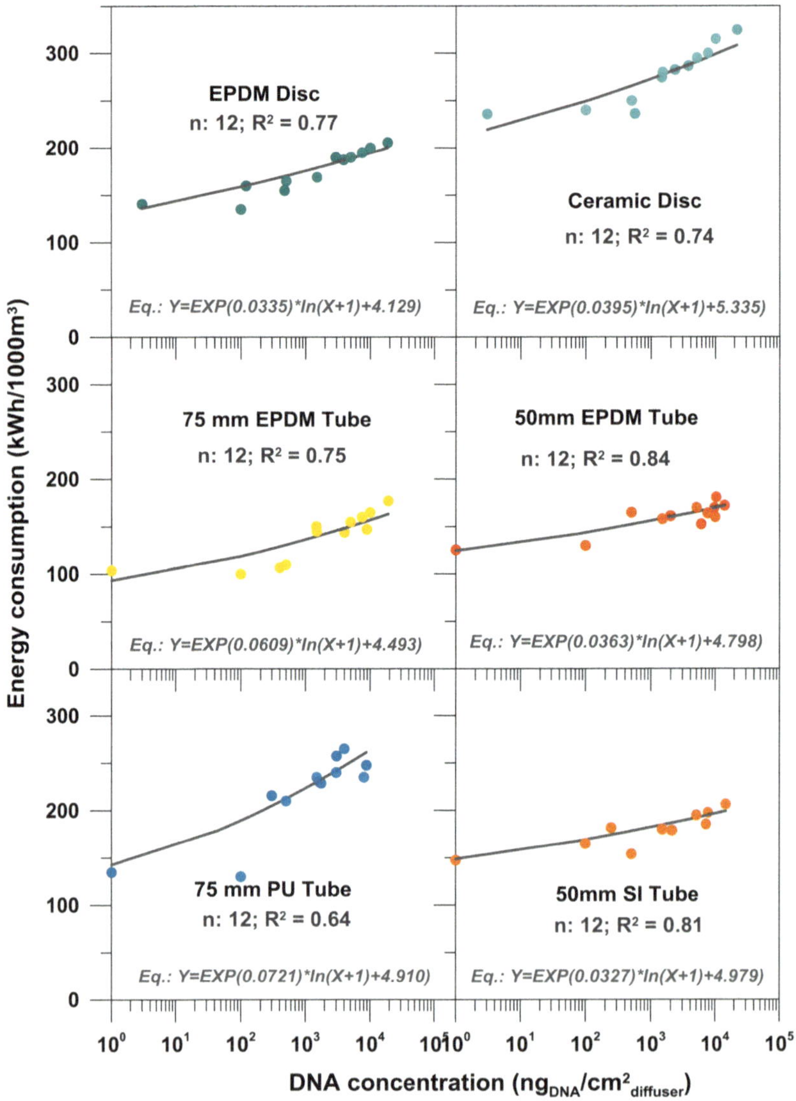

Figure 6.18 Correlation between energy consumption calculated with the aforementioned methodology and total biofilm DNA concentration for the six membranes. For each plot the coefficient of variation never exceeded the value reported in Figure 6.12.

created and applied, a set of conclusions could be drawn. Similar to the aforementioned disc diffusers, the rest also showed a sharp initial increase of biofilm growth, indicating that the initially rapid decrease in performance is due

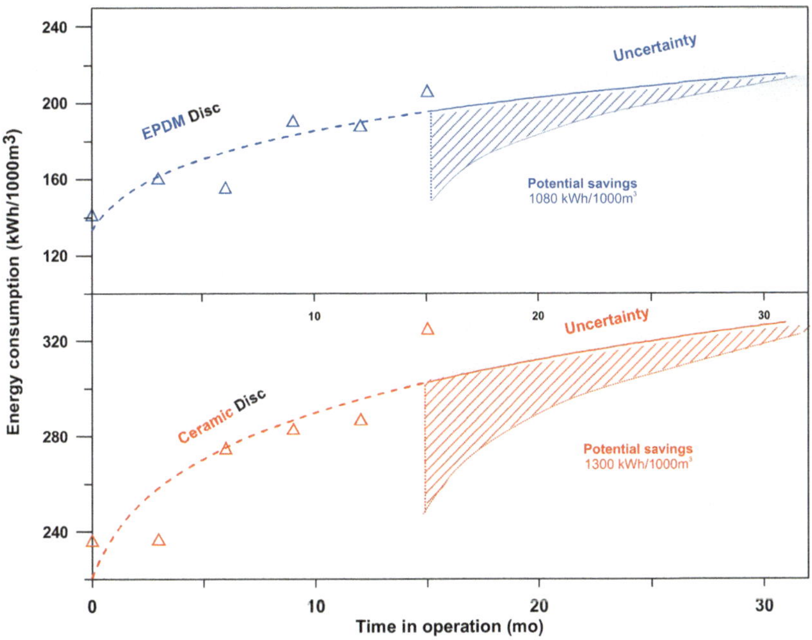

Figure 6.19 Energy consumption of the EPDM and ceramic disc diffuser over the 15 month period and the corresponding energy projections and estimated potential savings during the period from 15 to 32 months.

mainly to an increase in biofilm growth, while a more moderate increase was observed from 15 to 32 mo. Note that the average increase after 36 months resembles that experienced at 15 months for those diffusers installed in the higher-rate process (0.41 ± 0.05), indicating that high-rate processes may contribute to shortening the diffuser life expectancy by two- to three-fold due to the effect of microbial activity (biological aging).

Once fouling has caused a decline in diffuser efficiency, the efficacy of cleaning methods will vary and can be considered site-specific. Standardized techniques to project the efficiency of cleaning methods are not yet well-documented or established, though some studies have shown a capacity to restore efficiency by more than 75% in relative terms (Frey & Thonhauser, 2004; Kaliman *et al.*, 2008; Rosso & Shaw, 2015; Rosso & Stenstrom, 2006a, b; Umur, 2014; Wagner *et al.*, 2007). Therefore, in order to show relevance to forecasting energy demands of aeration diffusers, for our presentation here we theorized and estimated the potential benefits when cleaning strategies were applied after a period of 15 months in operation. Following the current literature and experience, a theoretical recovery of 75% of the initial diffuser performance after each cleaning episode was assumed.

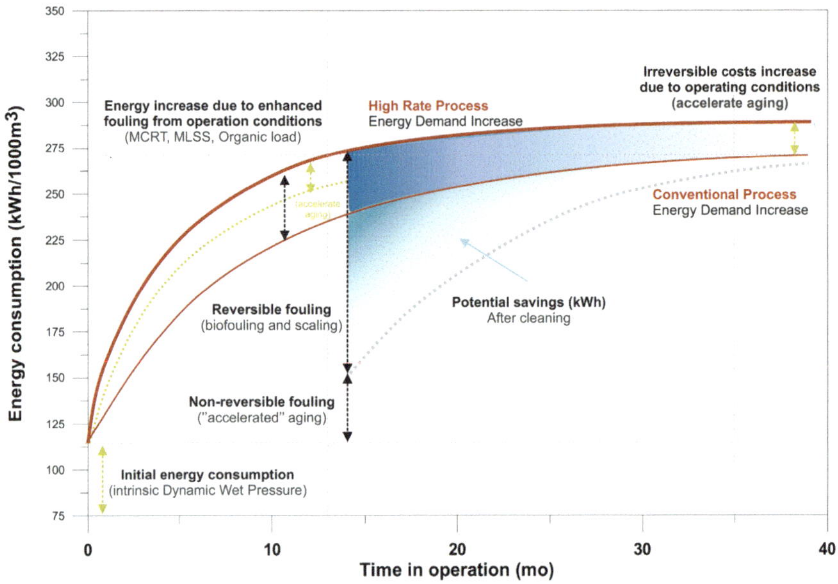

Figure 6.20 Hypothetical representation of the operating conditions (high-rate process vs. conventional system) impact on aeration diffuser fouling and concurrently on the energy demand (summary figure designed after analysing the project results).

Further research should provide a benchmark comparison for the different potential cleaning methods and corresponding efficiencies, together with the overall price. The incorporation of this information in the analysis would certainly improve future cost projections and decision-making in aeration systems selection and operation. For example, using this methodology, the estimated cost savings for a two-year forecast considering the same rate of fouling for a plant of 74,500 m^3 d^{-1} (or 20 MGD) would represent a savings for EPDM and ceramic discs of approximately 110,000 and 90,000 USD (assuming an energy price of \$0.1/kWh), respectively (Figure 6.20), whereas other diffusers show a wide range of potential savings, ranging from 53,000 to 200,000 USD (Garrido-Baserba *et al.*, 2017).

6.6 REFERENCES

Aberley R. C., Rattray G. B. and Dougas P. P. (1974). Air diffusion unit. *Journal of Water Pollution Control Federation*, **46**, 895–910.

Åmand L., Olsson G. and Carlsson B. (2013). Aeration control – a review. *Water Science and Technology*, **67**, 2374–2398.

Aymerich I., Rieger L., Sobhani R., Rosso D. and Corominas L. (2015). The difference between energy consumption and energy cost: modelling energy tariff structures for water resource recovery facilities. *Water Research*, **81**, 113–123.

Benedetti L., Prat P., Nopens I., Poch M., Poch M., Turon C., De Baets B. and Comas J. (2009). A new rule generation method to develop a decision support system for integrated management at river basin scale. *Water Science and Technology*, **60**, 2035–2040.

Cadet C., Béteau J. F. and Hernandez S. C. (2004). Multicriteria control strategy for cost/quality compromise in wastewater treatment plants. *Control Engineering Practice*, **12**, 335–347.

Chao W., Lingzhan M., Jun H., Peifang W., Jin Q. and Shanshan D. (2014). The effect of flow velocity on the distribution and composition of extracellular polymeric substances in biofilms and the detachment mechanism of biofilms. *Water Science and Technology*, **69**(4), 825–832.

Ekman M., Björlenius B. and Andersson M. (2006). Control of the aeration volume in an activated sludge process using supervisory control strategies. *Water Research*, **40**, 1668–1676.

Emami N., Sobhani R. and Rosso D. (2018). Diurnal variations of the energy intensity and associated greenhouse gas emissions for activated sludge processes. *Water Science and Technology*, **77**, 1838–1850.

Frey W. and Thonhauser C. (2004). Clogging and cleaning of fine-pore membrane diffusers. *Water Science and Technology*, **50**(7), 69–77.

Garrido-Baserba M., Sobhani R., Asvapathanagul P., McCarthy G. W., Olson B. H., Odize V., Al-Omari A., Murthy S., Nifong A., Godwin J., Bott C. B., Stenstrom M. K., Shaw A. R. and Rosso D. (2017). Modelling the link amongst fine-pore diffuser fouling, oxygen transfer efficiency, and aeration energy intensity. *Water Research*, **111**, 127–139.

Garrido-Baserba M., Asvapathanagul P., Park H.-D., Kim T.-S., Baquero-Rodriguez G. A., Olson B. H. and Rosso D. (2018). Impact of fouling on the decline of aeration efficiency under different operational conditions at WRRFs. *Science of the Total Environment*, **639**, 248–257.

Gernaey K., Jeppsson U., Vanrolleghem P. and Copp J. (2014). *Benchmarking of Control Strategies for Wastewater Treatment Plants (Scientific and Technical Report no 23)*, IWA Publishing, London, UK.

Gillot S., Capela-Marsal S., Roustan M. and Héduit A. (2005). Predicting oxygen transfer of fine bubble diffused aeration systems – model issued from dimensional analysis. *Water Research*, **39**, 1379–1387.

Graff Zivin J. S., Kotchen M. J. and Mansur E. T. (2014). Spatial and temporal heterogeneity of marginal emissions: implications for electric cars and other electricity-shifting policies. *Journal of Economic Behavior and Organization*, **107**, 248–268.

Griffiths-Sattenspiel B., Wilson W. and Hastings M. (2009). *The Carbon Footprint of Water*. A River Network Report, Portland, Oregon, USA.

Guerrero J., Guisasola A., Comas J., Rodríguez-Roda I. and Baeza J. A. (2012). Multi-criteria selection of optimum WWTP control setpoints based on microbiology-related failures, effluent quality and operating costs. *Chemical Engineering Journal*, **188**, 23–29.

Guo J., Fu X., Baquero G. A., Sobhani R., Nolasco D. A. and Rosso D. (2016). Trade-off between carbon emission and effluent quality of activated sludge processes under seasonal variations of wastewater temperature and mean cell retention time. *Science of the Total Environment*, **547**, 331–344.

Henze M., van Loosdrecht M. C., Ekama G. A. and Brdjanovic D. (2008). *Biological Wastewater Treatment: Principles, Modelling and Design*. IWA Publishing, London, UK.

Hertle A. R. H. (2015). Fine Bubble Diffused Air Aeration Systems. *Water Encyclopedia*. https://doi.org/10.1002/047147844X.ww98

Ingildsen P. and Olsson G. (2015). *Smart Water Utilities: Complexity Made Simple*. IWA Publishing. London, UK.

Jiang L.-M., Garrido-Baserba M., Nolasco D., Al-Omari A., DeClippeleir H., Murthy S. and Rosso D. (2017). Modelling oxygen transfer using dynamic alpha factors. *Water Research*, **124**, 139–148.

Kaliman A., Rosso D., Leu S. S. Y. and Stenstrom M. K. (2008). Fine-pore aeration diffusers: accelerated membrane ageing studies. *Water Research*, **42**, 467–475.

Kenway S. J., Lant P. A., Priestley A. and Daniels P. (2011). The connection between water and energy in cities: a review. *Water Science and Technology*, **63**, 1983–1990.

Kim Y. and Boyle W. C. (1993). Mechanisms of fouling in fine-pore diffuser aeration. *Journal of Environmental Engineering*, **119**, 1119–1138.

Leu S.-Y., Rosso D., Larson L. E. and Stenstrom M. K. (2009). Real-time aeration efficiency monitoring in the activated sludge process and methods to reduce energy consumption and operating costs. *Water Environment Research*, **81**, 2471–2481.

Mannina G., Ekama G., Caniani D., Cosenza A., Esposito G., Gori R., Garrido-Baserba M., Rosso D. and Olsson G. (2016). Greenhouse gases from wastewater treatment – a review of modelling tools. *Science of the Total Environment*, **551–552**, 254–270.

Metcalf & Eddy (2014). *Wastewater Engineering: Treatment and Resource Recovery*, 5th edn, McGraw-Hill Education, New York, NY, USA.

Olsson G. (2012). ICA and me – A subjective review. *Water Research*, **46**, 1585–1624.

Olsson G., Carlsson B., Comas J., Copp J., Gernaey K. V., Ingildsen P., Jeppsson U., Kim C., Rieger L., Rodríguez-Roda I., Steyer J. P., Takács I., Vanrolleghem P. A., Vargas A., Yuan Z. and Åmand L. (2014). Instrumentation, control and automation in wastewater – from London 1973 to Narbonne 2013. *Water Science and Technology*, **69**, 1373–1385.

Palmer J., Flint S. and Brooks J. (2007). Bacterial cell attachment, the beginning of a biofilm. *Journal of Industrial Microbiology and Biotechnology*, **34**, 577–88.

Reardon D. J. (1995). Turning down the power. *Civil Engineering*, **65**(8), 54–56.

Redmon D., Boyle W. C. and Ewing L. (1983). Oxygen transfer efficiency measurements in mixed liquor using off-gas techniques. *Journal of Water Pollution Control Federation*, **55**, 1338–1347.

Rieger L., Jones R. M., Dold P. L. and Bott C. B. (2014). Ammonia-based feedforward and feedback aeration control in activated sludge processes. *Water Environment Research*, **86**, 63–73.

Rieth M., Boyle W. and Ewing L. (1990). Effects of selected design parameters on the fouling characteristics of ceramic diffusers. *Research Journal of Water Pollution Control Federation*, **62**, 877–886.

Rosso D. and Shaw A. R. (2015). Framework for Energy Neutral Treatment for the 21st Century through Energy Efficient Aeration. Water Environ. Res. Found. (Project INFR2R12).

Rosso D. and Stenstrom M. K. (2005). Comparative economic analysis of the impacts of mean cell retention time and denitrification on aeration systems. *Water Research*, **39**, 3773–3780.

Rosso D. and Stenstrom M. (2006a). Economic implications of fine-pore diffuser aging. *Water Environment Research*, **78**, 810–815.

Rosso D. and Stenstrom M. K. (2006b). Surfactant effects on α-factors in aeration systems. *Water Research*, **40**, 1397–1404.

Rosso D. and Stenstrom M. K. (2007). Energy-saving benefits of denitrification. *Environmental Engineering: Applied Research and Practice*, **3**, 29–38.

Rosso D., Iranpour R. and Stenstrom M. K. (2005). Fifteen years of offgas transfer efficiency measurements on fine-pore aerators: key role of sludge age and normalized air flux. *Water Environment Research*, **77**, 266–273.

Rosso D., Stenstrom M. K. and Larson L. E. (2008). Aeration of large-scale municipal wastewater treatment plants: state of the art. *Water Science and Technology*, **57**, 973–8.

Samuelsson P., Halvarsson B. and Carlsson B. (2007). Cost-efficient operation of a denitrifying activated sludge process. *Water Research*, **41**, 2325–2332.

Sobhani R. (2011). Analysis of Diurnal Variations in Energy Footprint and Its Associated Carbon Emission for Water Supply and Reuse in Arid and Semi-Arid Areas. *PhD thesis*, University of California, Irvine, USA.

Sobhani R., Jiang L.-M., Chau A. M., Brown J., Torres E., Shao Y. J. and Rosso D. (2011). Energy footprint analysis of Orange County Sanitation District's operations. *Proceedings of Water Environment Federation*, **2011**, 3707–3714.

Spees K. and Lave L. (2008). Impacts of responsive load in PJM: load shifting and real time pricing. *The Energy Journal*, **29**(2), 101 121.

Stare A., Vrečko D., Hvala N. and Strmčnik S. (2007). Comparison of control strategies for nitrogen removal in an activated sludge process in terms of operating costs: a simulation study. *Water Research*, **41**, 2004–2014.

Umur S. (2014). *Aeration Systems Basics and Sanitaire Liquid Diffuser Cleaning System.* Xylem Inc., New York, USA.

Wagner M. R. and Pöpel H. J. (1998). Oxygen transfer and aeration efficiency - Influence of diffuser submergence, diffuser density, and blower type. *Water Science and Technology*, **38**(3), 1–6.

Wagner M. and von Hoessle R. (2004). Biological coating of EPDM-membranes of fine bubble diffusers. *Water Science and Technology*, **50**(7), 79–85.

Wagner M., Loock P. and Guenkel T. (2007). Design and operational aspects of modern fine bubble aeration systems. in: 10th IWA Specialised Conference: Design, Operation and Economics of Large Wastewater Treatment Plants, Vienna. *Preprints.*

WEF. (2009). *Energy Conservation in Water and Wastewater Treatment Facilities.* WEF Manual of Practice, WEF Press, New York, NY, USA.

Chapter 7

Modelling aeration and energy

Daniel Nolasco[1], Manel Garrido-Baserba[2], Gan Wang[3] and Diego Rosso[2]

[1]NOLASCO y Asociados
[2]University of California, Irvine
[3]PLAKeco Ltd.

7.1 WHY DYNAMIC MODELLING

Dynamic mathematical models are necessary for the description of the time-variant phenomena encountered in wastewater treatment processes. For decades, practitioners and researchers have recognized the many potential benefits derived from the development and use of dynamic models and the incorporation of control systems in wastewater treatment plants (Andrews *et al.*, 1974): increased performance, process intensification, reliability and stability, and a reduction in operational costs are among these benefits.

Aeration phenomena are dynamic. Diurnal changes in influent flow rate and contaminant concentrations, make the oxygen demand of incoming wastewater highly variable (see Sec. 6.3). When combined with time of use (TOU) electricity tariffs, which tend to charge more for electricity at those hours of the day when the highest demand for oxygen occurs, the importance of an adequate description of the aeration process through dynamic modelling becomes obvious (see Sec. 6.2). In activated sludge systems, the consumption of electrical energy for aeration represents an important part of the plant's operational costs.

Editor: Diego Rosso
doi: 10.2166/9781780407845_0215

Dynamic modelling of aeration systems becomes a key tool in the financial evaluation (cost-benefit analysis) of different control strategies. These control strategies must be combined with the selection of aeration equipment (blowers, diffusers and dissolved oxygen control systems), while taking into account process and effluent quality objectives. For these reasons, dynamic modelling of aeration systems, generally implemented within software programs (plant simulators), which simplify the computational aspects of mathematical modelling, has become an ubiquitous tool for the planning, costing, selection, design, control, and operation of water resource recovery facilities.

7.2 THE ART AND SCIENCE OF SPECIFYING ALPHA FACTORS

By now you are probably wondering why the book title does not contain the word 'alpha'. Much of the book has been dedicated to this factor that is the cause and (currently still) the solution to the quantification of process water effects on gas transfer. Some have proposed to retire it, however there is no viable alternative to replace it at this point. To front the need for defining useful ranges of α factors, we compiled here a set of recommendations for selecting α factors, based on our experience. Remember that these recommendations are guidelines based on our experience and should *never* be used in lieu of site-specific analyses, which should always be the approach of choice.

Let us spend a moment elaborating over some thermodynamic considerations on α, although this remark may appear philosophical to some:

It is fundamentally impossible to predict the α factor in wastewater

Backing up one step, we can consider the case of clean water and air. In this case we have a high degree of confidence in predicting the gas transfer coefficient k_La as a function of airflow and of the tank geometry. Even in the case when we add one contaminant, we can measure and predict k_La as a function of airflow, tank geometry, and the concentration of the known contaminant. Already in the 1970s, Zlokarnik (1980) had systematically studied the individual effect of different surface-active agents (whether they be aliphatic alcohols, inorganic salts, or amphiphilic organics) and identified α factors as a function of contaminant concentrations for both mechanical and diffused aeration (in cases when α was larger than 1.0 it was referred to as enhancement factor). However, the insurmountable problem surfaces when facing multiple contaminants or a solution of unknown composition. In fact, the interaction between the different contaminants creates a level of uncertainty that cannot be resolved unless the composition and the interaction between these contaminants are understood (the non-ideal activity of this solution would be a measure of entropy). The lack of knowledge on our thermodynamic system is equivalent to entropy, thus the prediction of gas transfer coefficients (hence, of α) in solutions of unknown composition and activity would be tantamount to a decrease in entropy, which would violate the fundamental laws of thermodynamics (Dr Nernst and Dr Thomson, best known as Lord Kelvin, would not be too pleased with us taking this liberty). Therefore, the only approach left to us is to define ranges of α within which we cannot know (and *should not* attempt to identify) a single exact value for α.

As seen in Table 7.1, the α factor increases from the conventional carbonaceous removal process (ranging from 0.2 to 0.4) to the NDN process (ranging from 0.6 to 0.9), and reaching the maximum efficiency in the DEMON™ process (close to 0.95). The improvement is related to the increase of MCRT operation time, and for the DEMON™ process is also related to the change in nature of the oxygen demand applied to the process. In fact, as remarked above, even though the COD applied to the deammonification process is similar in concentration to that of another processes, the nature of the COD is different as it comes from sludge processing. This sidestream COD is not expected to contain surfactants, which would depress oxygen transfer (e.g., soap and detergents), therefore able to maintain a high α under process conditions.

7.2.1 Alpha accuracy during design and specification of aeration systems

Bubble size and distribution strongly influence oxygen transfer. The combination of diffuser type (i.e., material, geometry), air supply (i.e., blowers, compressors) and the receiving load characteristics (e.g., surfactants concentration, TSS) will determine the range or average in transfer efficiency (i.e., α values or SOTE) expected from a certain type of diffuser in given conditions or site-specific characteristics. Diffusers are designed to release a certain bubble size distribution, but forecasting a homogeneous air column in bubble diameter is only true in clean water conditions. The eventual size of the bubble during operation will largely vary depending on wastewater density, temperature, turbulence, bubble coalescence, wastewater constituents, hydrodynamics, etc. Providing an exact α value for specific types of diffusers can lead, then, to false or lower expected performances. This section provides a methodology to evaluate and quantify the expected α in a given scenario in order to improve design calculations. Furthermore, some values obtained in earlier studies are also provided.

For example, when WRRFs need to install or upgrade the existing aeration system one of the initials steps entails a request of data and design parameters from manufacturers of candidate diffusers. The selected manufacturers will provide the design team detailed guidelines to adhere to in order to successfully design the aeration system. However, the efficiency parameters provided and warranted by manufacturers are based on historical experience in other similar installations and on their internal experience (Rosso *et al.*, 2013). A methodology based on an *in situ* column method can be used to evaluate diffusers (Fig. 7.1). This testing has to be performed in clean and process water (wastewater) at the

Table 7.1 Differences in daily average α values for different processes according their associated MCRTs. Note the dynamic nature of aeration efficiency parameters.

Source	Carbonaceous Removal Only			Nitrifying	NDN			DEMON™
	Literature	DC Water	Sydney Water	Literature	Literature		IWRD	HRSD
MCRT	1–8	6.5–8	7–8	10–20	10–15	15–30	7–11	N/A
$\langle\alpha\rangle$	0.2–0.5	0.4–0.6	0.4–0.7	0.5–0.65	0.2–0.5	0.6–0.9	0.55–0.7	0.8–0.95
SOTE (αSOTE/Z) %/ft *%/m*	0.5–0.9 0.9–1.1 *0.15–0.33*	1.4–1.5 *0.42–0.45*	1.2–1.7 *0.36–0.51*	0.9–1.45 0.9–1.3 *0.27–0.44*	1.1–1.6 *0.33–0.48*	0.9–1.3 *0.27–0.42*	1.4–1.5 *0.42–0.45*	*Not available*

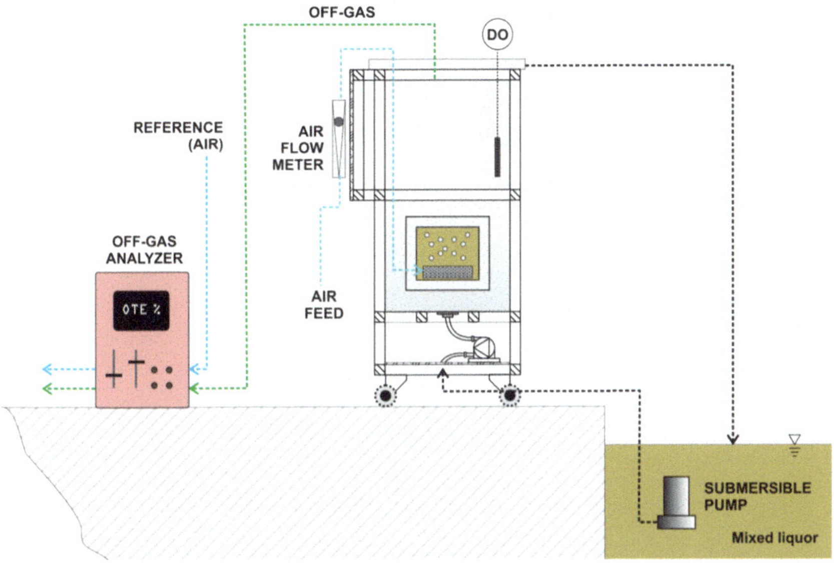

Figure 7.1 Schematic of on-site column testing for process water. Adapted from Rosso *et al.* (2013).

treatment facility. The obtained results will help to bypass the uncertainty caused by extrapolating results from different wastewaters and projects. Moreover, an extended fouling test in the plant's wastewater can be carried out within the design timing, providing the designers with site-specific aeration performance data.

To illustrate some results obtained with this method, Table 7.2 presents the collected measurements of seven different types of membrane diffusers utilized by two different WRRFs during a period of one year. The table summarizes the observed effects of different diffuser types to show the advantages and disadvantages of selecting a type of diffuser for a specific scenario. It can be seen that all diffusers seem to experience a decrease in α or αF through time until reaching what appears a stable value around 0.4 and 0.5. In this case, the highest αF values could be found in EPDM type membranes, where values around 0.8 were observed. On the other hand, although some diffusers initially experienced the lowest α, their progressive decrease is much lower in comparison to other diffusers, and even after one year they still maintain values similar to the initial values.

Moreover, Table 7.2 shows that αSOTE/Z is influenced by the diffuser membrane typology (Z stands for diffuser submergence, in metres). Diffuser geometry (i.e., tubes versus discs) has shown significant effect. All tube diffusers have shown αSOTE/Z values ranging from 1.3 to 1.9%/ft (4.27–6.23%/m), whereas discs presented lower values ranging from 0.65 to 0.90%/ft (2.13–2.95%/m). Note that discs usually have a surface area half that of a tube.

Table 7.2 Site-specific α reported in different diffusers types in pilot units.

1 Year Test No D.O. Cotnrol	**2″ EPDM Tube**	**2″ Silicone Tube**	**9″ EPDM Disc**	**9″ Ceramic Disc**	**3″ EPDM Tube**	**3″ PU Tube**
$\alpha_{initial}$	0.65	0.5	0.8	0.45	0.8	0.75
αF	0.5	0.4	0.5	0.35	0.45	0.5
(αSOTE)/Z [%/ft]	1.3–1.7	1.1–1.5	1.1–1.6	0.65–0.9	1.3–2.1	0.8–1.6
Fouling ($\alpha F/\alpha_{new}$)	0.37 ± 0.05	0.28 ± 0.06	0.27 ± 0.04	0.17 ± 0.06	0.15 ± 0.02	0.2 ± 0.1
DWP$_{initial}$ (inch of water)	19 ± 2	27 ± 3	38 ± 5	8 ± 1	22 ± 3	19 ± 1
Expected P$_w$ increase	20%	25%	45%	35%	50%	70%
Visual Examination After 15 months						
DNA Concentration (ng/μL)	10,000 ± 3000	14,500 ± 8000	7500 ± 8000	8000 ± 1000	20,000 ± 5000	6500 ± 1000

Therefore, the diffuser density for tubes is much higher than that of discs and may be key in defining efficiency patterns.

7.2.1.1 Dangers of using an α for coarse-bubbles when using jets or turbines

The α factor for a coarse-bubble diffuser may not be applied to a jet or turbine because of the different flow regimes of the bubbles generated by these different technologies. Coarse bubbles travel rapidly throughout their life, while fine bubbles generated by jets or turbines travel fast in the jet zone but exhibit behavior consistent with that of fine bubbles from fine-pore diffusers once they spread in a plume.

In the case of an aerobic digester, this is even more evident. In this case, the thickness of the sludge promotes bubble coalescence (Baquero-Rodríguez *et al.*, 2018). Since the thick sludge is a non-Newtonian shear thinning fluid, in order for the bubbles to experience less hydrodynamic resistance (i.e., lower viscosity) they had to travel faster, which in a gravitational field can only occur when the bubbles are larger. Therefore, bubble coalescence is promoted by the sludge and the gas transfer that in clean water would be satisfactory for the small bubbles produced by turbines and jets becomes severely impaired in process water. In other words, the k_La in clean water experiences a dramatic decrease in 'a', which is proportional to $6/d_B$, where d_B is the bubble diameter. The net result is a much lower oxygen transfer rate than expected, due to the overestimation of the α factor. The oxygen deficiency in the process resulted in very low or undetectable DO values and the inability for the process to meet target treatment values.

A design based on this fallacy faced potential scenarios spanning from design amendments to litigation. A simple consultation with a third-party expert can correct this design flaw at its inception.

7.2.2 Daily dynamics and α

The oxygen transfer efficiency (OTE, %) varies during wastewater treatment because of the effects of wastewater contaminants. Previous reports show that α factors fluctuate through daily cycles (Leu *et al.*, 2009) and seasonal periods (Libra *et al.*, 2005) depending on process conditions such as organic loading rates, dissolved oxygen (DO), temperature, etc. The variations of the α factor substantially influence the overall oxygen transfer performance of the aeration system. Many authors have agreed on the impact of surfactants on bubble diameter and the reduction of the overall liquid-side volumetric mass transfer coefficient k_La (Capela *et al.*, 2002; Gillot & Héduit, 2008; Rosso & Stenstrom, 2006; Stenstrom & Gilbert, 1981; Wagner & Pöpel, 1998). However, results differ according to the nature of the surfactants, their concentration, and the observed system (Rosso & Stenstrom, 2005a; Sardeing *et al.*, 2006). Routine and more practical measurements in WRRFs, such as mixed liquor suspended solids (MLSS) and chemical oxygen demand (COD), have been targeted as potential indicators of the α factor variations.

The reduction of gas transfer efficiency with increased COD concentration is unfortunate, since the period of greatest aeration rate occurs at the lowest aeration efficiency and tends to overlap with the highest energy and power demand charges (Aymerich *et al.*, 2015), and, in many cases, with the highest cost of electricity. Therefore, because high organic loadings drive α to its lowest value when the oxygen (i.e., energy) demand is the highest, the implications of dynamic α on energy costs should be of the utmost concern. This phenomenon occurs consistently in almost all WRRFs, where the influent characteristics fluctuate daily, weekly, and monthly. On an average daily basis, the fluctuation between an insufficiency and an excess of OTR results in process deficiency in both meeting treatment objectives and containing energy costs. The large changes in α factor experienced are of maximum interest for process designers, as they directly determine the process efficiency and concurrently the aeration power requirements. However, to date, the use of constant α factors are the most common practice in process design and modelling, thus promoting inaccuracies in the oxygen provision and the constant need for model recalibration to compensate for a lack in structure (Amaral *et al.*, 2016; Plano *et al.*, 2011). Only recently have process models incorporated the ability to specify time-variant α factors.

Although many plants have efficiently deployed automatic control to address daily dynamics, many of them did not accurately account for the compounding effect on air requirements of decreasing OTE and α factors at peak loading, which led to the inefficient design of blowers and the concurrent incapacity to optimize the controllers (Amerlinck *et al.*, 2016; Schraa *et al.*, 2017). Therefore, many plants systematically discharge more air than required, as they simply tend to use a conservative blower specification to compensate for the low α factors during peak loading periods. During those peak periods, the depression in α can be corrected by higher airflow rates, but maintaining these airflow rates during longer periods to avoid the risk of complying can have cost and efficiency consequences. The state of the art in aeration design and modelling using commercially available simulators based on the Activated Sludge Model (ASM) family (Henze, 2000; Henze *et al.*, 1987) is to rely on user-inputs of OTE curve-fitting parameters and constant or scheduled α factors. The OTE curves typically only correct for air flux, diffuser density, and submergence (Barker & Dold, 1997; DHI, 2014 ; Snowling, 2014). Because both OTE and α vary with load, both of them exhibit a dynamic nature that should be accounted for in a truly dynamic model. Numerous attempts at developing descriptive or predictive relationships for α have been published (Gillot *et al.*, 2005; Gillot & Héduit, 2008; Plano *et al.*, 2011; Rosso & Stenstrom, 2005b; Wagner & Pöpel, 1998). However, none of these attempts have been completely successful in describing the true dynamic nature of α.

Hence, the use of a dynamic α becomes necessary to address the real process dynamics. Designers need to be aware of the varying α factors to provide process flexibility and respond to temporal loading fluctuations. Aeration systems should be designed with the needed 'turn up' and 'turn down' capacity to accommodate such changes. Knowing and predicting dynamic α would also limit the uncertainty in predicting the blower size and number at peak, the time-dependent blower power demand, the cumulative aeration energy footprint, and ultimately the aeration system capital and operating costs.

7.2.2.1 Developing a model to narrow the uncertainty on α

The introduction of a dynamic α in the model structure, calibrated using full-scale data sets, increases the prediction quality for airflow rate and all of the operating parameters derived therefrom. A model was developed based on the results of two WRRFs were 24 h off-gas test were carried out, providing during a 24 h period the corresponding SOTE, α, and relevant parameters. Once the model was developed, it was calibrated and validated using data from two water reclamation plants operating at similar conditions.

The 24 h experiment results carried at one of the WRRFs provided supporting evidence for previous observations made by Leu *et al.* (2009), where the organic loading (i.e., COD) depressed oxygen transfer and α factors (Figure 7.2). The results from both 24 h experiments are presented in Figure 7.2 showing the COD

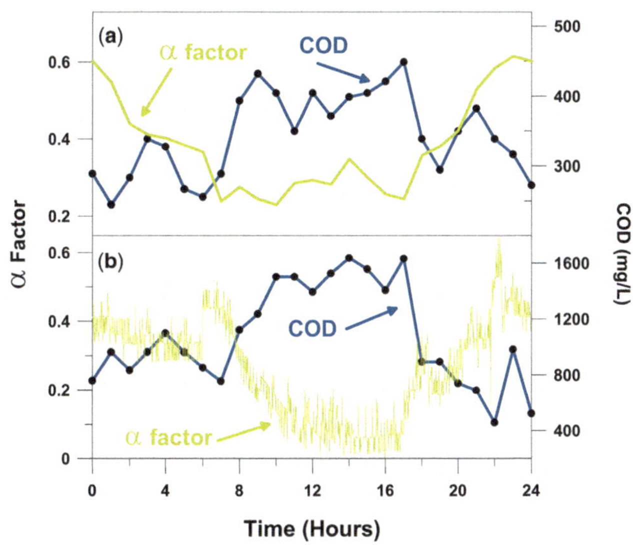

Figure 7.2 Influent load and α factor (α) during a 24 hr cycle at two different WRRFs. The α factors were measured by off-gas tests every hour (a) and minute (b). (a) Data from Leu *et al.* (2009) (WRRF1); (b) data from WRRF3, June 2015.

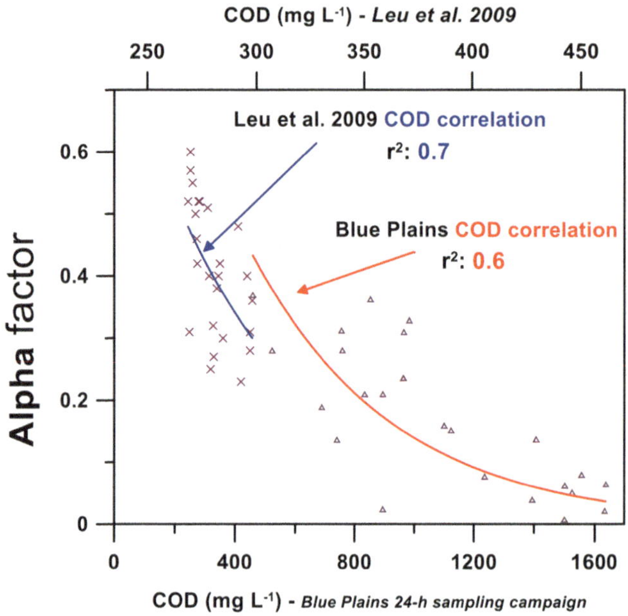

Figure 7.3 Alpha factor (α) and total COD. The negative correlation of α as a function of the local COD (i.e., measured at the same point where the off-gas is measured) over a diurnal cycle (data from Leu *et al.*, 2009; WRRF3). Aggregate data from both studies is represented with red dotted line ($r^2 = 0.7$).

and the α factor negatively correlated as a function of the influent diurnal COD plant loading. The full-scale sampling campaigns presented specific correlations for each WRRF between daily patterns in COD and the aeration efficiency parameters (i.e., OTE or α factors).

A negative correlation between α and COD for each treatment site was identified (Figure 7.3), as previously observed (Germain *et al.*, 2007; Steinmetz, 1996). The observed correlation between total COD and α factor has a coefficient of determination of 0.7, which is high taking into account that both WRFFs studied are using two different aeration systems. Whereas the data collected by Leu *et al.* (2009) was obtained from an aeration basin where fine-pore diffusers were installed, at the other WRRF a coarse-bubble aeration system was recently deployed. This therefore strengthens the idea that an overall correlation to assess the α variation as function of COD is possible by an appropriate monitoring of the upcoming organic load, independently of the aeration system installed.

7.2.2.2 Mathematical formulations of dynamic α

After more than a decade of investigation, an exponential fit was found by the authors to describe a correlation between α and process variables. The relationship between α and plant load is described in this model with the following fit:

$$\alpha = A \cdot e^{-K_{\alpha} \cdot [\mathrm{COD}]} \tag{7.1}$$

where A and k_{α} are fitting parameters (function adjustment by residuals minimization), and COD is total Chemical Oxygen Demand (COD) concentration ($mg\ l^{-1}$).

Following Equation (7.1), the model is able to incorporate the effects of wastewater contaminants (α factor) on the variation in aeration efficiency. Thus, the model structure includes the dynamic recalculation of air flow as a function of the total COD and the actual (not estimated) oxygen transfer. The α curves designed with the aforementioned objective were calibrated using published (Leu *et al.*, 2009) and newly measured time-dependent data for α and COD in the wastewater. The calibration set of this model adds parameters for A and k_{α} (fitting parameter for function residual minimization).

The modelling results showed that using a high α implies the delivery of 27.3% less air flux than the actual requirements, which may result in an underperformance of the treatment plant. By choosing a conservative constant α of 0.35 would increase the air wastage by 29.4%, in comparison to the actual air flux requirements due to the delivery of an excess of oxygen to compensate the low α value. An α value of 0.35 would imply strong interference by contaminants hindering oxygen transfer. On the other hand, expecting better oxygen transfer than actual can also create negative consequences. A generous α value of 0.80 is also shown in Figure 7.4, since it was the commonly chosen value for fine-pore diffusers in the early diffuser installations (Metcalf & Eddy, 1979).

The more important comparison is that between the best-fit constant α (=0.50, i.e., the midpoint in the possible range $\alpha = [0,..,1]$) and the dynamic α. For this specific scenario, the best fit corresponded to an α of 0.50, which provided remarkably close results to the actual air flux requirements, with a range of +6.6% and −5.9% for air wastage and under delivery, respectively. On the other hand, the dynamic α provided the closest values to the actual flux requirements (+3.0% wastage and −2.9% deficit). To quantify the quality of the model fit, the sum of squared error (SSE) was calculated for the hourly airflow rate (as in Gonzalez & Kamrani, 2001). The SSE for the 24 h period under study was 7.04, while the corresponding value between the best fit ($\alpha = 0.50$) and the actual airflow rate was 9.1. Therefore, it can be observed that the dynamic model improves the predictive capacity of the best fit (constant) by approximately 23% (based on SSE).

Note that other formulations for α as functions of process variables exist. Gillot & Héduit (2008) introduced the equivalent contact time (ECT) as an auxiliary variable for a prediction of α. Knowing the MCRT, the $k_{L}a$, and the airflow rate, one can predict the α factor using their rational procedure. The ECT combines in one variable the effect that multiple factors (airflow rate, diffuser submergence,

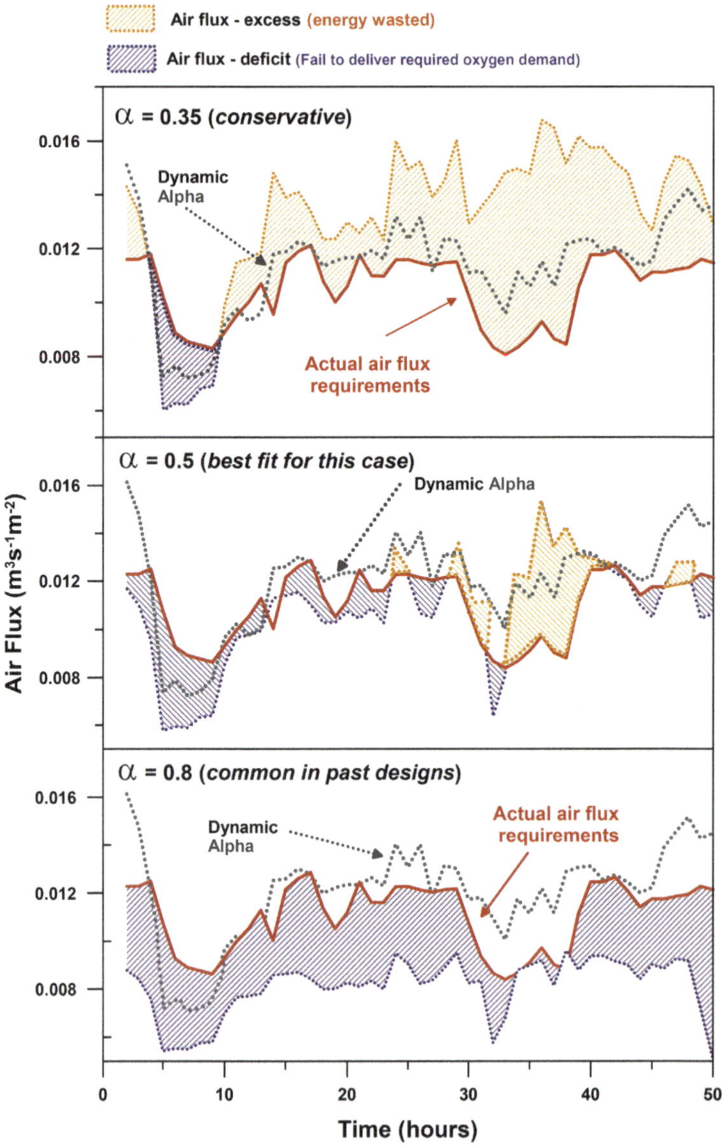

Figure 7.4 Comparison of the constant α factor values (conservative = 0.35; best fit = 0.50; typical of old designs = 0.80) against the dynamic α, for the case of WRRF2. The red line represents the actual air requirements. The purple areas below the red line represent the oxygen deficit, while the orange areas are the oxygen wastage. For the cases of $\alpha = 0.35$; 0.5; 0.8; and dynamic the sum of squared error (SSE) are respectively 23.9; 9.1; 37.6 and 7.0, demonstrating that the dynamic α best represents the actual data.

horizontal flow) exhibit on mass transfer. This work was based on an elegant dimensional analysis (published earlier in Gillot *et al.*, 2005), however one limitation of this report is the application to plants with very low F/M ratio, and α factors were in the range 0.44–0.98.

7.2.3 Modelling α profiles

Whether the process is conceptualized as a well mixed reactor, a series of well mixed reactors, or a plug-flow reactor, α factors still remain necessary for process design. This section describe how to model different types of processes and how the common assumptions are made by experts in modelling practice, by providing ranges of α profiles along the process trains.

Two major factors determine the quality of the effluent leaving a biochemical treatment process: the kinetics of the reactions occurring within the system and the characteristics of the reactor within which these reactions occur (Grady *et al.*, 2011). Therefore, an understanding of the basic behaviour of the aeration tanks (the reactors) is of paramount importance for modelling. The aeration tank can behave as a batch reactor, a continuous stirred tank reactor (CSTR), a plug-flow reactor (PFR) or a number of CSTRs in series. Batch reactors are initially charged with a feed containing the reactants, sealed, and then the required reactions take place for a predetermined period of time. An example of the application of batch reactors in activated sludge systems is the sequencing batch reactor (SBR). A considerable number of variations of SBRs that depart from the 'sealed' condition for operation can be encountered in the industry (e.g., continuously fed SBRs). CSTRs, also known as completely mixed reactors, have a continuous feed stream (influent) and an exit stream (effluent). For the mixing to be assumed 'perfect', a powerful mixer operated at a very high speed would be required. The energy from submerged aeration could also be used for mixing purposes. Needless to say, perfect mixing conditions do not exist in reality, but for modelling purposes, they could be assumed to be ideal.

Under perfect mixing conditions, any reactant carried into the reactor by the feed is dispersed evenly throughout the reactor in no time. Therefore, samples taken from all parts of the reactor have the same composition. When operating under a fixed liquid volume (V_r), the effluent flow rate (Q_e) equals the influent flow rate (Q_o). Under all operating conditions, the effluent composition (whether particulate matter, X or soluble matter S) is equal to the suspended solids concentration (X_e). The same thing happens to the concentration of soluble material (S). The plug-flow reactor (PFR) operates like a simple tube.

A feed containing the reactants is fed continuously to the reactor inlet, while the effluent is removed from the outlet, located at the other extreme of the tube. In an ideal PFR, the flow pattern inside has uniform velocity and concentration in the radial direction at any point along the length of the reactor, with no longitudinal (axial) dispersion. For this reason, the concentrations of reactants and products in a PFR vary along the reactor length (unlike a CSTR, which has uniform concentrations throughout the reactor). The operation of an ideal PFR equals that of an infinite number of CSTRs of infinitesimal volume placed in series. However, for practical purposes, four CSTRs in series closely resemble the behaviour of an ideal PFR. Simulators of wastewater treatment plants generally use ideal CSTR tanks as unit processes (aerated/un-aerated tanks for activated sludge, digesters, etc.). The user can select how many CSTRs are needed to adequately resemble the behaviour of the plant in study: from an ideal CSTR to a PFR, passing through some intermediate conditions. Since ideal conditions (whether CSTR or PFR) do not exist in reality, the modeller must decide which configuration (i.e., number and disposition of CSTRs) better resembles the field conditions of the treatment system (e.g., activated sludge reactors or lagoons) to be modelled.

For a given total reactor volume in a facility, the number of CSTR reactors in series will have a considerable influence on the rate of removal of contaminants, as well as on the distribution of oxygen requirements, energy consumption, etc. Therefore, the determination of the most approximate number of CSTRs needed to simulate adequately a plant is important to adjust the modelling exercise as close as possible to reality.

Some cases are relatively easy, as the one shown in Figure 7.5, where the aeration tanks of this conventional activated sludge plant, visually and operationally, resemble two CSTRs in parallel. They are both aerated with fine-pore diffusers uniformly distributed (which can be found from as-built drawings not shown herein), the surface of each tank resembles a square, and nothing on the surface of the mixed liquor gives the impression that mixing is not reasonably uniform. Therefore, simulating these tanks as two CSTRs operating in parallel seems to be the obvious answer (see Figure 7.6).

In surface aeration systems, like the one shown in Figure 7.7, in spite of not having dividing walls, each surface aerator generates a quasi-CSTR condition. Thereby, the system can be modelled like a series of CSTR tanks, one for each surface aerator. Long tanks with dividing walls, like the one shown in Figure 7.8, can also be simulated as a series of CSTRs (each division marks the beginning and end of a CSTR).

Other systems are not so easy to define. For instance, if the tank shown in Figure 7.8 did not have any divisions, how could it be modelled? As one CSTR, two CSTRs in series, three…? Different modellers have different methods to answer this question. The most accurate way to define the hydraulic behaviour of a tank and determine the number of CSTRs in series that best resemble such behaviour would be to perform tracer studies (i.e., Rhodamine WT) or some further analysis (De Clercq *et al.*, 1999). However, tracer studies can be expensive and time-consuming. A much simpler approach would be to use rules of thumb. For instance, we can assume that tanks that are considerably longer than wide can be divided into a series of CSTRs, assuming that every six widths there is a

Figure 7.5 Example of aeration tanks with fine pore diffusers presenting uniformly distributed mixing. These tanks have a geometry that resembles two CSTRs in series. In this case, the mixing is guaranteed by full floor coverage with membrane panels. The top image shows one tank in operation, the bottom image is of an adjacent tank after diffuser maintenance and operating at shallow depth with secondary effluent before resuming operations, to protect the diffusers from sunlight.

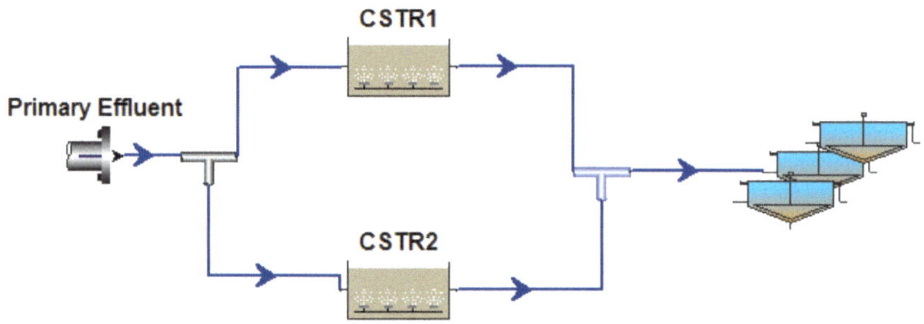

Figure 7.6 Model representation of the full-scale aeration tanks using fine pore diffusers.

CSTR. It has to be noted that starting from four CSTRs in series and beyond, the behaviour of the system resembles that of an ideal PFR. Therefore, having more than four or five CSTRs in series increases the computational demand of the model, without much practical improvement in the modelling accuracy.

Since not all tank lengths can be divided exactly by six widths, some common sense and modelling experience must be exercised. In fact, one can argue that there could be only as many CSTRs as there is ability to independently

Figure 7.7 Surface aeration systems without having dividing walls and generating almost perfect CSTR conditions.

Figure 7.8 Long tanks with dividing walls can be simulated as a series of CSTRs (each division marks the beginning and end of a CSTR).

control zone mixing, which in a tank with multiple aeration grids in series would correspond to the number of diffuser grids. There are relatively simple tools available for modellers that offer an intermediate analysis between the tracer studies and the above rule-of-thumb. One of these tools is an EXCEL spreadsheet to calculate the equivalent number of CSTR of tanks in series according to tank geometry, flow and aeration is proposed in: https://www.iwapublishing.com/sites/default/files/CopyofWaterWiki3Estimationofnumberoftanksinseries.xlsx.

Another method of dividing into a number of CSTR is by each aeration grid. Since many of today's aeration basins are equipped with a tapered diffuser density, this method makes some sense and actually accounts for the actual existing diffuser density.

Once the number of CSTRs in series is determined, the following task needed for aeration modelling is the determination of the α value for each reactor (for field testing, see Chapter 2). However, when the plant being modelled has not been designed yet and no field-testing can be performed, some sort of guidance must be provided for the initial estimation of these factors. Experience shows that α factors increases along the length of a PFR: smaller at the upstream end of the tank and higher at the downstream end (Fisher & Boyle, 1999). As the bacterial population progressively degrades the quantity of contaminants (i.e., surfactants, COD, etc.) along the tank, the oxygen transfer efficiency increase (together with the α factor) as the effect of contaminants in mass transfer is decreased.

In a three CSTR in series system, the first tank may have α values ranging from 0.25 to 0.35, which would increase along the tanks to α values in the proximity of 0.60 to 0.70 in the last CSTR. The influence of fouling (F) will depend on the longevity of the diffusers: being close to 1.0 for relatively new diffusers (all along the series of tanks), down to values starting at 0.7 in the first CSTR up to 0.95 in the last CSTR for tanks that have been operating for one year. Therefore, the product of α factor for three CSTRs in series with less than a year of operation may range from 0.20 to 0.30 for the first tank, increasing up to 0.50–0.70 in the last tank.

Having unaerated tanks located upstream of the system, as is the case for anaerobic/anoxic selectors or BNR systems, the α factor value of the first aerated tank immediately downstream of the first unaerated zone will be considerably higher than that of a first CSTR that receives fresh primary effluent. This is due to the reduction of readily biodegradable organic matter (rbCOD) occurring in unaerated zones. Less rbCOD entering the aerated tank means less surfactants available to reduce α factor, as well as less available organics to cause diffuser (Groves *et al.*, 1992).

7.3 ENERGY MODELLING OF AERATION PERFORMANCE

7.3.1 SOTE curves vs. the 2%/ft assumption

As discussed, the oxygen transfer efficiency is defined as the ratio between the mass of oxygen transferred as DO into the wastewater over the mass of oxygen delivered (through the blowers) into the tank, and it is generally expressed as percentage. *Ceteris paribus*, the higher this percentage is, the more efficient the aeration system, that is, the less energy will be required to achieve the desired treatment level. Therefore, it is important to understand the factors that influence OTE and how to model it dynamically.

Under standard conditions the SOTE (S stands for standard) is a function of the aeration system configuration: diffuser type, diffuser density (DD: total area of diffusers over total aeration tank area), diffuser depth in the tank, and airflow rate per diffuser (Figures 7.9 and 7.10). The understanding and quantification of the impact of different operational parameters can help design of the aeration system, but the incorporation of this knowledge into simulator tools can improve current modelling capabilities.

For example, diffuser manufacturers test their diffusers to determine SOTE under various operational and installation conditions. An example of such curves is presented in the Figure 7.9. For a specific type of diffuser, these types of curves

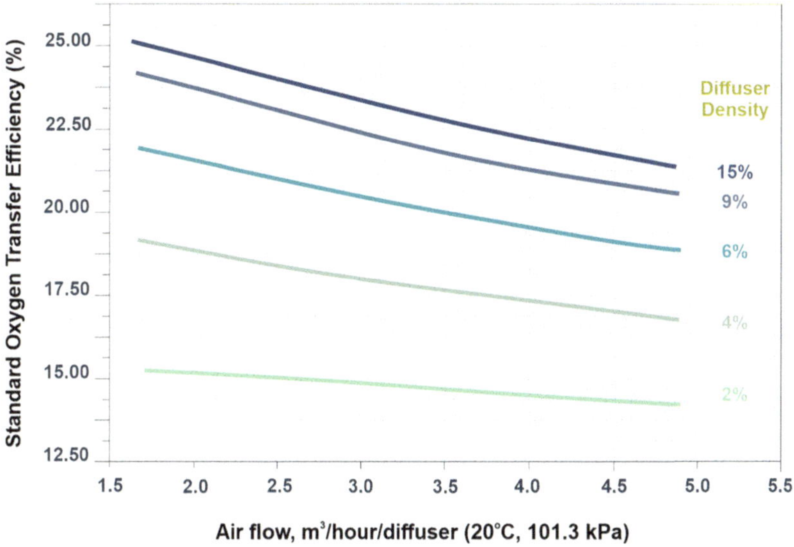

Figure 7.9 Example of performance curves provided by manufacturers to determine SOTE under various operational conditions. The diffuser density is the ratio of the total diffusing area to the tank floor.

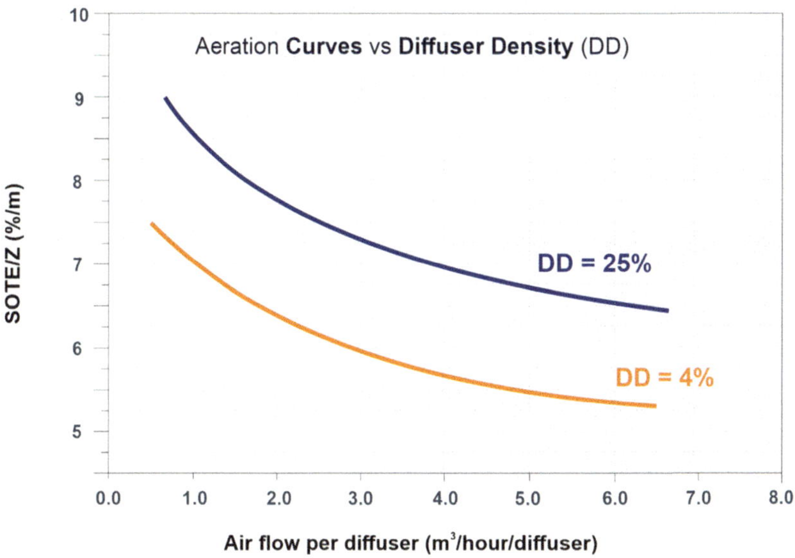

Figure 7.10 An example of the way a dynamic simulator represents the variation of SOTE, in this case expressed as a % of SOTE/m of diffuser submergence can be seen in the following picture.

can be entered in dynamic simulators, so as these parameters vary, the SOTE (%) is adjusted automatically by the simulator, without the need of modeller input.

Figure 7.10 illustrates another example of the way a dynamic simulator represents the variation of SOTE, in this case expressed as a % of SOTE/Z. While the dots represent the values taken from the manufacturer's data, the solid lines show the model curves adjusted to match them. As can be seen from the figure, current simulators allow for quite a reasonable adjustment.

When designing new plants, SOTE curves can be combined with biological models that predict hourly, diurnal, and seasonal oxygen demand to determine the optimal number of diffusers and the maximum, minimum and average airflow rates to be expected. Airflow rates can then be used to determine the number and type of blowers to be used. Energy requirements can also be calculated, thereby estimating costs of electricity by combining such requirements with electricity tariffs (see Chapter 6). Electricity tariffs can also be entered into today's dynamic simulators.

The detrimental effect on transfer efficiency due to the coalescence and formation of larger bubbles (reducing the specific area) when increasing the airflow rate can also be included in current simulators. Without modelling tools, the reduction in SOTE as the airflow rate per diffuser increases is next to impossible to model. A simplification used in spreadsheet models is to assume that SOTE is constant (assuming diffuser density and depth are kept constant). When assuming SOTE to be constant, the k_La

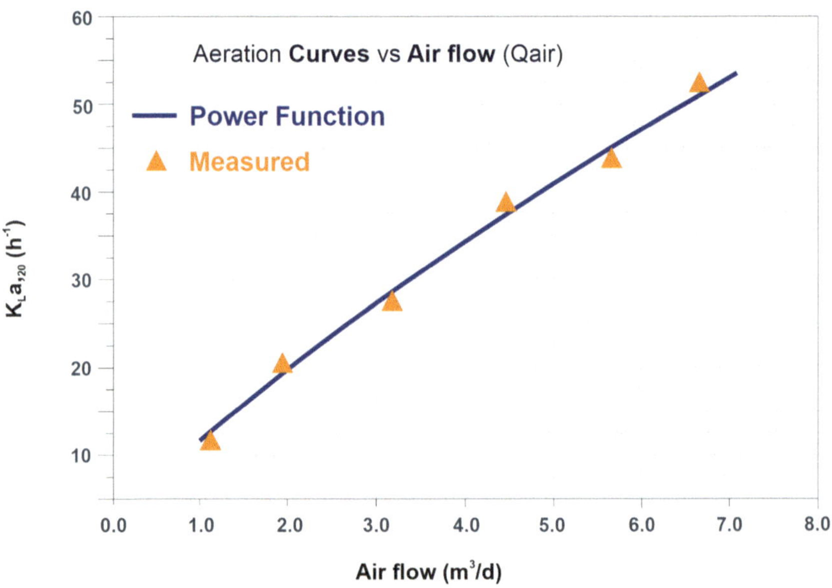

Figure 7.11 An example of the way a dynamic simulator represents the variation in oxygen mass transfer depending on the airflow (Q_{air}).

value increases in direct proportion to the mass of oxygen fed to the tank. However, since SOTE is not constant (as shown in the previous figure), SOTE diminishes as Q_{air} (or the mass of oxygen pumped into the tank, which is proportional to Q_{air}) increases for a fixed number of diffusers. The actual variation of k_La as a function of airflow rate (Q_{air}) looks as proposed in Figure 7.11.

Assuming SOTE to be constant may lead to inadequate aeration system design and operation. This is particularly the case when implementing automatic control of dissolved oxygen. To adjust to the hourly variations in oxygen demand, automatic DO control requires frequent adjustments of Q_{air} to maintain the DO at the setpoint (see Chapter 5). Since the number of diffusers in the tank does not change, variations in Q_{air} into the aeration tank imply variations in Q_{air} per diffuser, which modifies SOTE and thereby, the OTE under field operating conditions.

Modelling DO control with constant and dynamic SOTE

To present an example of this situation, a dynamic simulator was calibrated to an existing wastewater treatment plant. The plant is an activated sludge plant with preliminary treatment, primary clarification, conventional activated sludge and solids processing consistent of primary and secondary sludge thickening, anaerobic digestion and biosolids dewatering. The layout of the plant, as implemented in the simulator, is presented in Figure 7.12 and Figure 7.13.

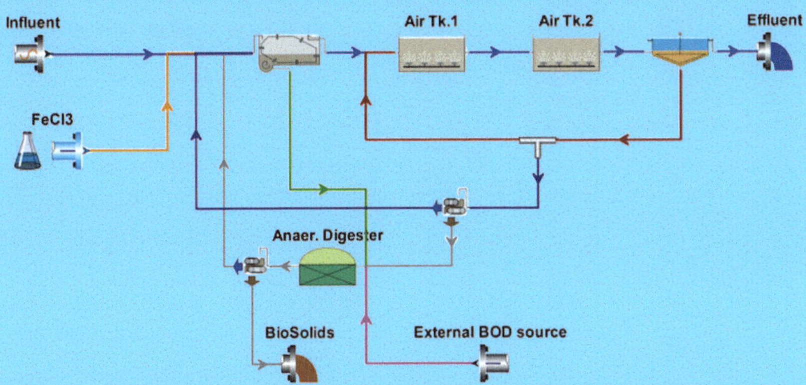

Figure 7.12 The layout of a typical WRRF implemented in a simulator.

Step 1) WRRF Characterization. Average influent characteristics of this example municipal treatment plant:

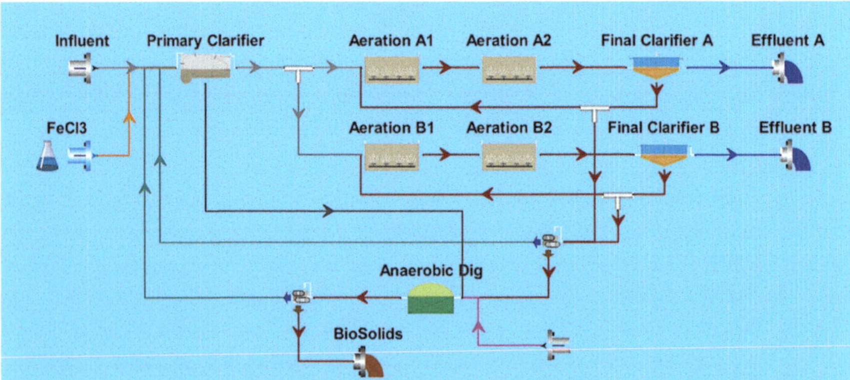

Figure 7.13 Another example of layout of a typical WRRF implemented in a simulator. Those model files implementing parallel treatment trains can be used to run at different conditions in both trains to compare them side-by-side (i.e., effect of DO control, constant or variable SOTE, effluent quality, etc.).

Daily average influent conditions

Flow	11200.10	m3/d
Total COD	644.01	mg/L
Total Kjeldahl Nitrogen	40.00	mgN/L
Total P	12.00	mgP/L
ISS Influent	81.00	mg/L
Volatile suspended solids	124.76	mg/L
Total suspended solids	206.60	mg/L
Total Carbonaceous BOD	377.72	mg/L
pH	7.00	

Although the influent data given above is daily average data, the actual influent data inserted into the model is variant on an hourly basis allowing a true dynamic analysis to be completed (Fig. 7.14).

Unit process dimensions are summarized in the following tables.

Name: Dec. Prim.	Type: Ideal primary settling tank	
Volume:	2268.00	m3
Area:	567.00	m2
Depth:	4.00	m

Name: Air Tk.1	Type: Bioreactor	
Volume:	2931.00	m3
Area:	488.50	m2
Depth:	6.00	m
Diffuser coverage:	14.50	%
Number of diffusers:	1728	
Diffuser unit area:	0.0410	m2

Name: Air Tk.2	Type: Bioreactor	
Volume:	2931.00	m3
Area:	488.50	m2
Depth:	6.00	m
Diffuser coverage:	14.50	%
Number of diffusers:	1728	
Diffuser unit area:	0.0410	m2

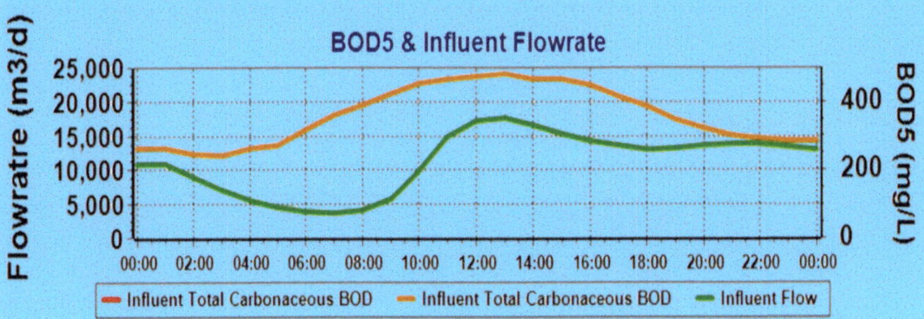

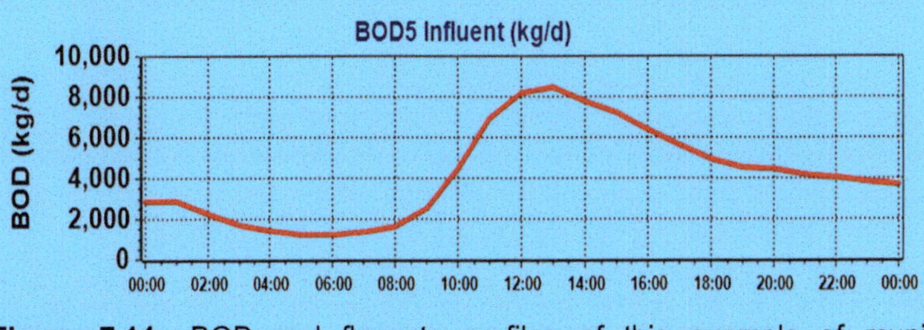

Figure 7.14 BOD and flowrate profiles of this example of municipal treatment plant.

Name: Dec. Sec.	Type: Ideal clarifier	
Volume:	4924.00	m3
Area:	1231.00	m2
Depth:	4.00	m

The plant is operated at approximately 12 day solids retention time and achieves an effluent quality (daily average values) as shown in the following table.

Flow	11192.10 m3/d
Ammonia N	0.09 mgN/L
Nitrate N	28.59 mgN/L
Nitrite N	0.02 mgN/L
Filtered TKN	1.95 mgN/L
Total N	32.27 mgN/L
Total P	9.73 mgP/L
Total suspended solids	25.47 mg/L
Total COD	50.20 mg/L
Total Carbonaceous BOD	9.24 mg/L
pH	7.00

The plant can be operated with or without automatic DO control. When no control is applied, the airflow rate in the tanks is constant. Under those conditions, the SOTE = 39% and remains constant at that value, since the airflow rate per diffuser is constant (Figure 7.15; Left). Once the automatic DO control is applied and a DO setpoint of 2 mg/L is used, the airflow rate is changed by the control system at each simulation interval to match the oxygen demand (Figure 7.15; Right). Under those conditions, SOTE should vary, decreasing as the airflow rate per diffuser increases. However, if the model used assumes SOTE to remain constant at 39%, independent of any changes in airflow rate, as in the following figures, the model will estimate different SOTE, OTE, Q_{air} required, and energy demand. A comparison of the situation with and without varying SOTE is presented in the following figures. A one-week dynamic simulation is used to evaluate weekly energy and cost implications.

Note that the graphs show two lines for power demand: one continuous on top, which represents the total power demand at the plant and one dashed below, which represents the net use of power after subtracting the power generated by the co-generation facility running with digester biogas.

In this example, the difference in terms of energy consumption is in the order of 10%, which may not result in a major change of aeration equipment (i.e., blowers, diffusers, piping and control system). However, when it comes to energy costs, 10% may represent an important difference to be considered. Also, the difference observed may vary under different equipment and plant operating conditions. This difference cannot be anticipated. Therefore, when available, SOTE variation with airflow rate per diffuser should be modelled realistically, as many dynamic simulators allow.

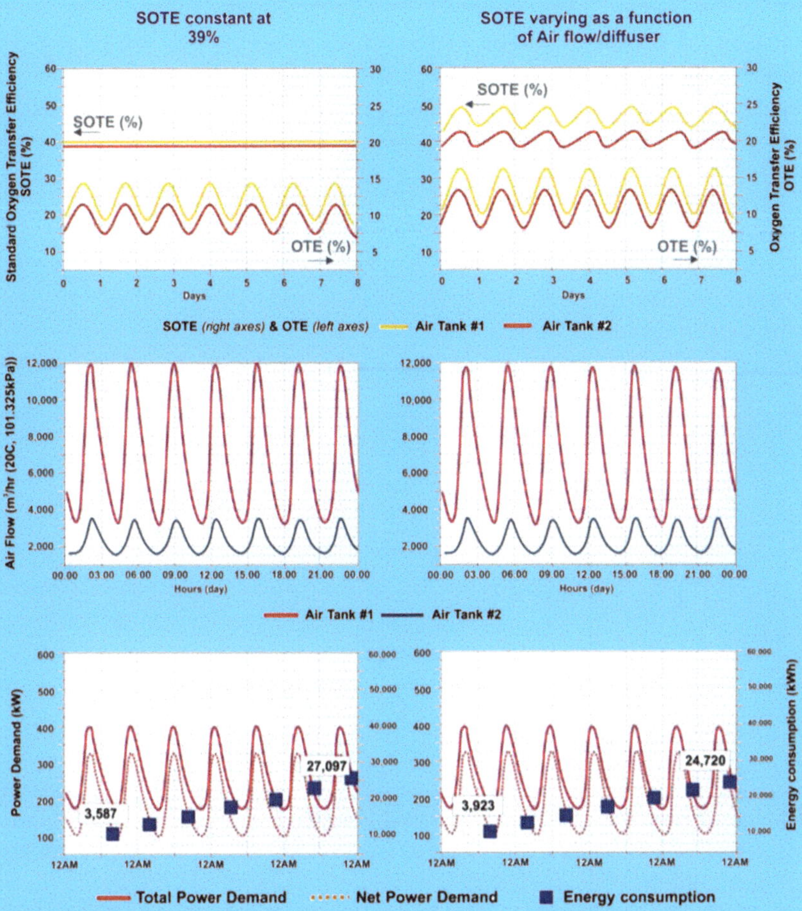

Figure 7.15 Comparison of the simulation with constant SOTE (39%) and with varying SOTE.

7.3.2 Constant vs. dynamic α

As indicated earlier in this chapter, the oxygen transfer efficiency (OTE) is affected by the α factor: the lower the α factor, the lower OTE, which in turn demands higher airflow rates and energy expenditures to transfer the same amount of oxygen to the mixed liquor. Alpha varies along the day as a function of COD. In general, under normal operating conditions, the COD in the influent is higher at the same time the influent flow rate is higher and, in many cases as well, at approximately the same time when electricity costs the most (under TOU tariff structures). In other words, worst-case conditions of low OTE, high demand and high-energy cost typically overlap at approximately the same time of day (Fig. 7.16). This is the case of the municipal treatment plant used as an example (Fig. 7.17).

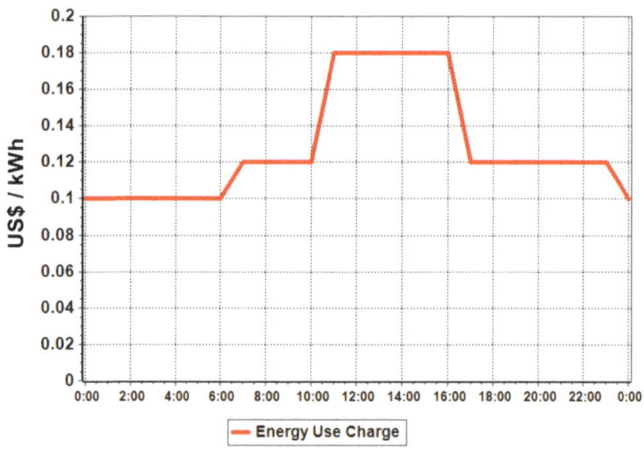

Figure 7.16 Representation of a tariff structure.

Modelling DO control with constant and time-dependent α

Under the current tariff structure, which this plant operates, the electrical energy cost varies depending on the time of use, as shown in the following figure.

A typical αF factor (the product of alpha factor times fouling factor) variation is presented in the following figure. This variation is used in the model of the example plant and compared vs. the case of assuming a constant α factor equal to the average of such variation ($\alpha F = 0.36$ and 0.44 for the first tank and second tank, respectively), shown in the following figure.

Note the impact of the variable αF factor on the oxygen transfer efficiency (OTE). The relatively minor variation in OTE observed when a constant αF

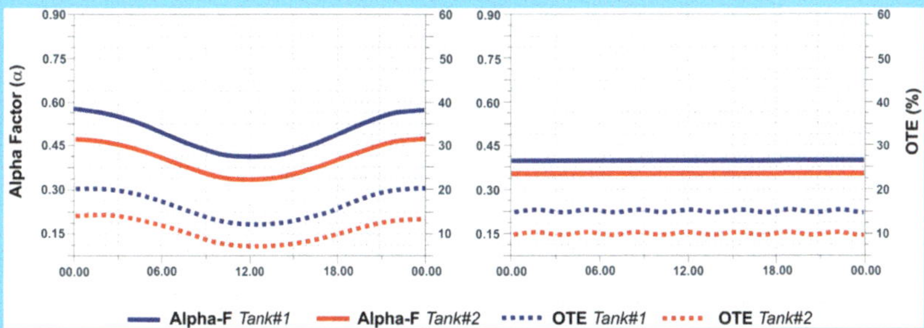

Figure 7.17 Simulation comparison between dynamic and constant α factor.

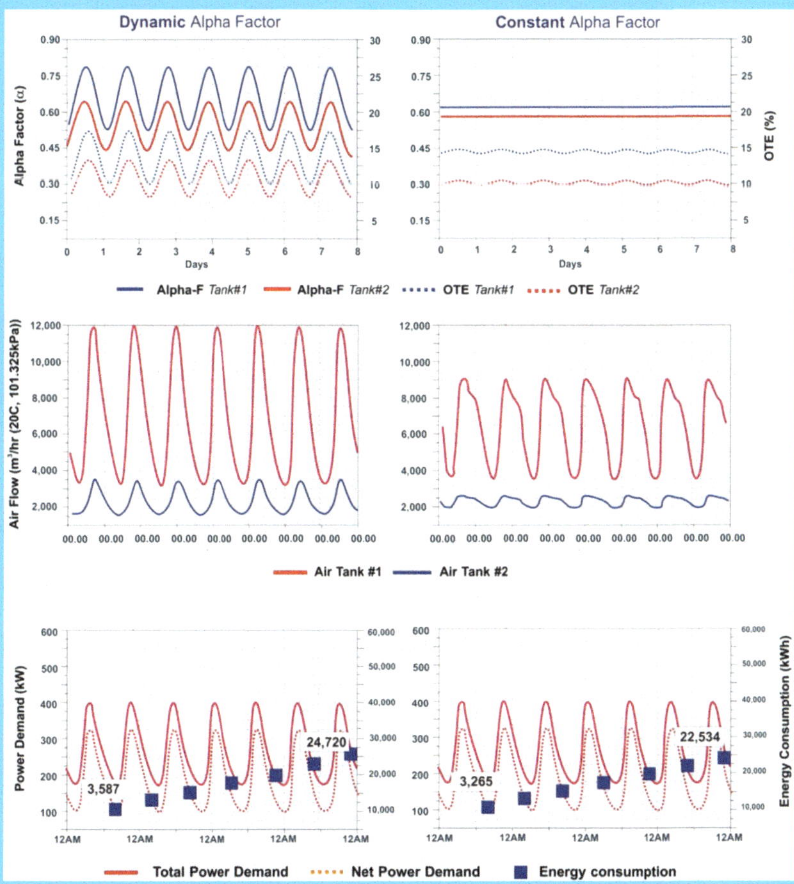

Figure 7.18 Comparison of simulations using a dynamic vs. constant α factor.

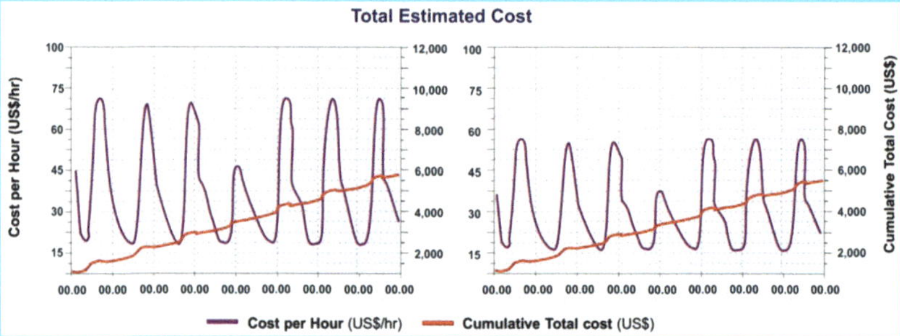

Figure 7.19 Comparison of simulated costs using a dynamic vs. constant α factor.

factor is used is due to the variation in airflow rate needed to control dissolved oxygen at a setpoint of 2 mg/L (used in both graphs).

The following figures (Figs. 7.17, 7.18 and 7.19) show dynamic simulations (7 days) comparing the use of a dynamic α factor (diurnal variation of αF, as shown in the previous graphs) vs. constant αF value at the average values (also shown in the previous picture).

When dynamic α (actually, αF) variations are taken into account, the electricity costs of the plant increase by US$2200/week, approximately (US$115,000/year). Perhaps the most important impact is related to the sizing of the blowers. For dynamic α conditions, the peak airflow rate reaches 12,000 m^3/hour. While for constant α conditions, the peak airflow rate is less than 9000 m^3/hour.

7.3.3 Finding the optimal DO setpoint

Selecting an adequate DO setpoint in an automatic aeration control system influences energy consumption (Fig. 7.20). In general, most treatment plants select ~2 mg/L as the DO control setpoint. However, if mixing conditions in the tank are adequate (e.g., no dead zones in tank corners and uniform distribution of dissolved oxygen throughout the basin, i.e., no relevant DO gradients) and such mixing conditions can be maintained at lower DO setpoints, it may be optimal to operate at a DO setpoint lower than 2 mg/L. Dynamic modelling is recommended to simulate these conditions and quantify potential energy savings, while confirming that there is no negative impact on effluent quality. Also, mapping the DO throughout the tank volume to verify that mixing is effectively good should be done (physically or via CFD).

For illustrative purposes a DO setpoint of 1 mg/L is tested in the example plant simulated in the previous sections. Alpha factor and SOTE conditions were modelled dynamically to make the simulation as realistic as possible. The results of such simulations are summarized in the following figures. Setpoint

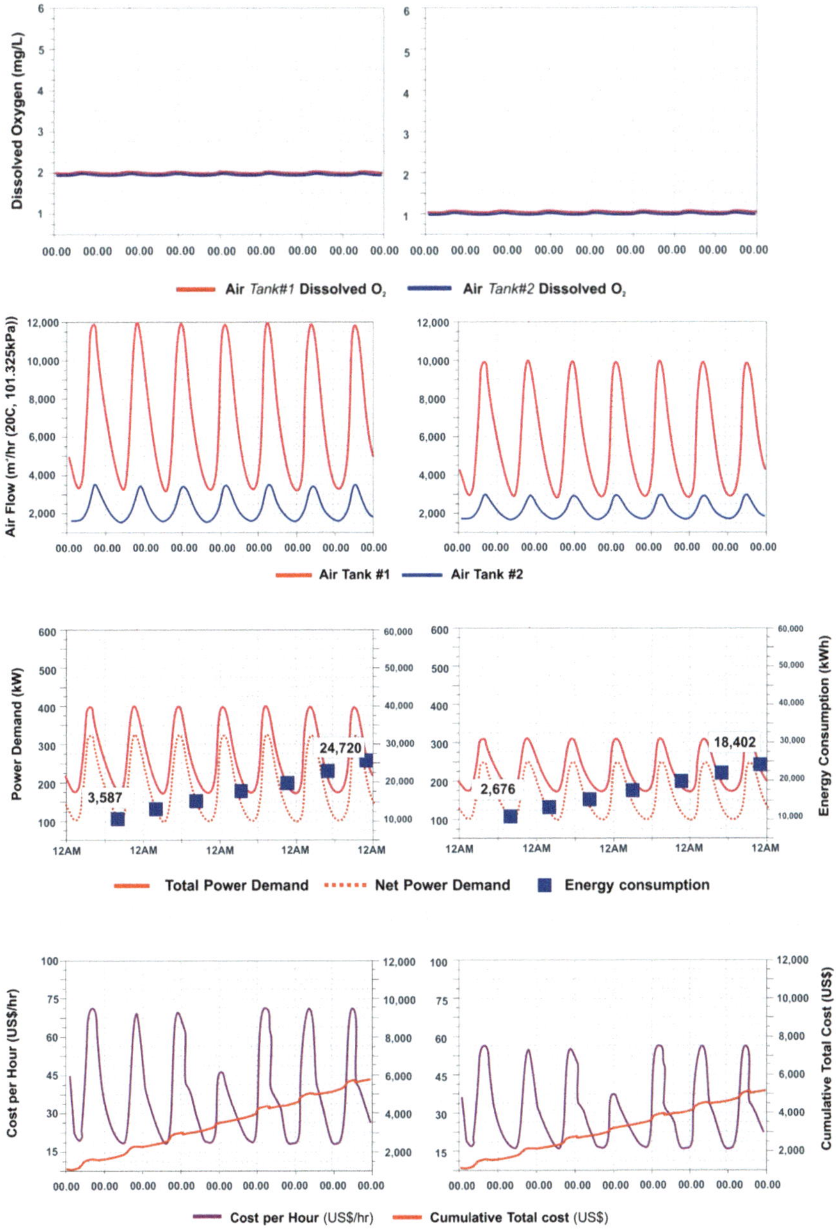

Figure 7.20 Comparison of the simulation results between the two DO strategies.

conditions of 2 mg/L and 1 mg/L DO are presented on the left and right side, respectively.

As expected, airflow rates for DO setpoint = 1 mg/L are considerably lower than those for 2 mg/L. Since k_La is directly proportional to airflow rate, the system with a lower DO setpoint requires less Q_{air}, and thereby, less energy to blow air. Thus, lower airflow rate releases smaller bubbles, hence an increase in contact surface area and the concurrent oxygen transfer efficiency.

Total energy consumption over the week of simulation is lower for the 1 mg/L setpoint (18,400 kWh) than for the 2 mg/L setpoint (24,720 kWh). This results in cost savings of approximately US$900/week (from US $6100/week for 2 mg/L setpoint to US$5200/week for 1 mg/L setpoint, approximately). This represents savings of US$47,000/year without any additional capital expenditures (CapEx) or operation and maintenance (O&M) expenditures (OpEx).

The next important question is: was the effluent quality impacted? The following comparison set out in Table 7.3 should answer this obvious question.

From an effluent quality standpoint, at this facility switching from 2 mg O_2/L to 1 mg O_2/L has no impact whatsoever on effluent water quality.

7.3.3.1 Further optimization options evaluated through dynamic modelling of aeration system

Numerous other design, control, and operational options can be tested and evaluated from an effluent quality and CapEx and OpEx with a dynamic simulator. For example:

Table 7.3 Average effluent quality values obtained with usual and reduced DO setpoint levels.

	DO Setpoint = 2 mg/l	DO Setpoint = 1 mg/l
Flow	13,085.00 m^3/d	13,085.00 m^3/d
Ammonia N	0.09 mgN/L	0.11 mgN/L
Nitrate N	27.15 mgN/L	24.92 mgN/L
Nitrite N	0.03 mgN/L	0.04 mgN/L
Filtered TKN	1.98 mgN/L	1.98 mgN/L
Total N	30.88 mgN/L	28.66 mgN/L
Total P	9.61 mP/L	9.61 mP/L
Total Suspended Solids	25.54 mg/L	25.59 mg/L
Total COD	50.99 mg/L	51.05 mg/L
Total Carbonaceous BOD	9.49 mg/L	9.55 mg/L
pH	7.00	7.00

(1) **Cleaning the diffusers** will reduce fouling, thereby increasing the fouling factor (F) which, as shown in [Equation #3] and [Equation #9] considerably reduces the Q_{air} needed to meet the target oxygen transfer rate (OTR). This in turn reduces energy consumption and OpEx at the facility. The change in F related to diffuser cleaning can be measured using oxygen transfer tests. Taking into account the cost of diffuser cleaning and energy costs, dynamic modelling can be used to determine the optimal frequency for cleaning, which is usually assumed to be 1–2 years, but actually different at every plant (see Chapter 2), as discussed in Rosso and Stenstrom (2006).

(2) **Unaerating the first zone in the aeration tank**, turning it into an anaerobic/anoxic zone. If the system nitrifies (and continues to nitrify after turning off the air in the first zone), the unaerated zone will be anoxic (i.e., no O_2 present, just NO_x). This will have several advantages: an oxygen credit due to denitrification in this zone; increased nitrogen removal, leading to better effluent quality and better settling, since the first zone will act as a biological selector controlling the growth of filamentous organisms, also leading to more and better effluent quality; and a reduction in the amount of surfactants (generally expressed as readily available COD), which in turn increases the α factor in the second tank. If the system does not nitrify, the unaerated zone will be anaerobic (i.e., no O_2 present, nor NO_x present). This will also have advantages: potential biological phosphorus removal; better settling, since the first zone will also act as a biological selector; and a reduction in the amount of surfactants (i.e., higher α factor in the aerated tank).

(3) **Analysing the effect of plant elevation on aeration demand**: Owing to a reduction in the saturation concentration of oxygen at higher altitudes, activated sludge plants require considerably higher airflow rate to meet similar treatment demands than plants located at lower altitudes. In Chapter 2, the Ω factor, that is, the correction factor for atmospheric pressure (which is approximated by p_b/p_s, where p_b is ambient barometric pressure and p_s is standard pressure at mean sea level), is considerably lower at higher altitudes, thereby requiring a higher k_La to keep the same OTR. The higher the k_La, the higher the Q_{air} required, and thereby the higher the energy input into the aeration system.

(4) **Introducing CEPT:** Chemically enhanced primary treatment (CEPT) reduces the oxygen demand in the aeration system by chemically mediating the sedimentation of contaminants in the primary clarifiers. This in turn sends more organic-rich sludge to the anaerobic digesters, which in turn produces more biogas that can be used to generate more electricity in a co-generation system at the plant. This is done at the cost of chemicals (precipitant added) and more sludge/biosolids to be disposed of. A cost-benefit analysis can be done with a dynamic simulator. Care must be taken when implementing CEPT to ensure

sufficient readily degradable organic remain in the primary effluent for BNR plant to maintain biological nutrient removal without adding external supplemental carbon.

(5) **Modelling the advantages of co-generation:** Since most dynamic simulators can quantify energy (heat + electricity) from biogas, when running dynamic simulations for aeration system evaluation, results for other evaluations can be compiled simultaneously. Since biogas may not be sufficient to cover all electricity demand at the plant, electricity can be generated at that time of the day when the tariff is most expensive. Modelling dynamically can therefore provide the quantitative basis for the cost–benefit analysis for the acquisition of a co-generation system.

(6) **Evaluating the advantages of co-digestion:** Additional sources of biodegradable organics can be added to the anaerobic digesters to boost the production of biogas, thereby increasing energy co-generation. Ideally, the sources of biodegradable waste should be industries that produce highly concentrated and highly biodegradable liquid effluent (e.g., bottlers of carbonated drinks and dairy products, breweries, restaurants). Highly concentrated liquid wastes introduced into municipal plant anaerobic digesters have little impact on the hydraulic retention time, temperature, and mixing conditions within these reactors (three parameters that are key to adequate digestion performance). However, the increase in biogas production can be dramatic and, in some cases, sufficient to cover the energy requirements for the whole facility, making it independent from the electrical grid and even, in some cases, turning it into a net producer of renewable energy. Higher sludge production is expected from this type of operation, as well as higher oxygen demand in the aeration tanks (due mainly to more concentrated contaminants returning from the sludge treatment train: digester supernatant, sludge dewatering liquid stream, etc.). Dynamically modelling co-digestion of various types of liquid wastes can help evaluate the impact on the aeration system, determine the adequate waste stream to use, study the quantity and frequency of such discharges into the digesters, and estimate the additional cost in sludge/biosolids disposal.

7.3.3.2 Simulation of low-cost, high-return modifications to aeration system

Testing the options mentioned above with a dynamic simulator is possible. Several permutations and combinations could be used, making this exercise considerably time-intensive. The option of turning off the air in the first of the two reactors, while lowering the dissolved oxygen setpoint in the second

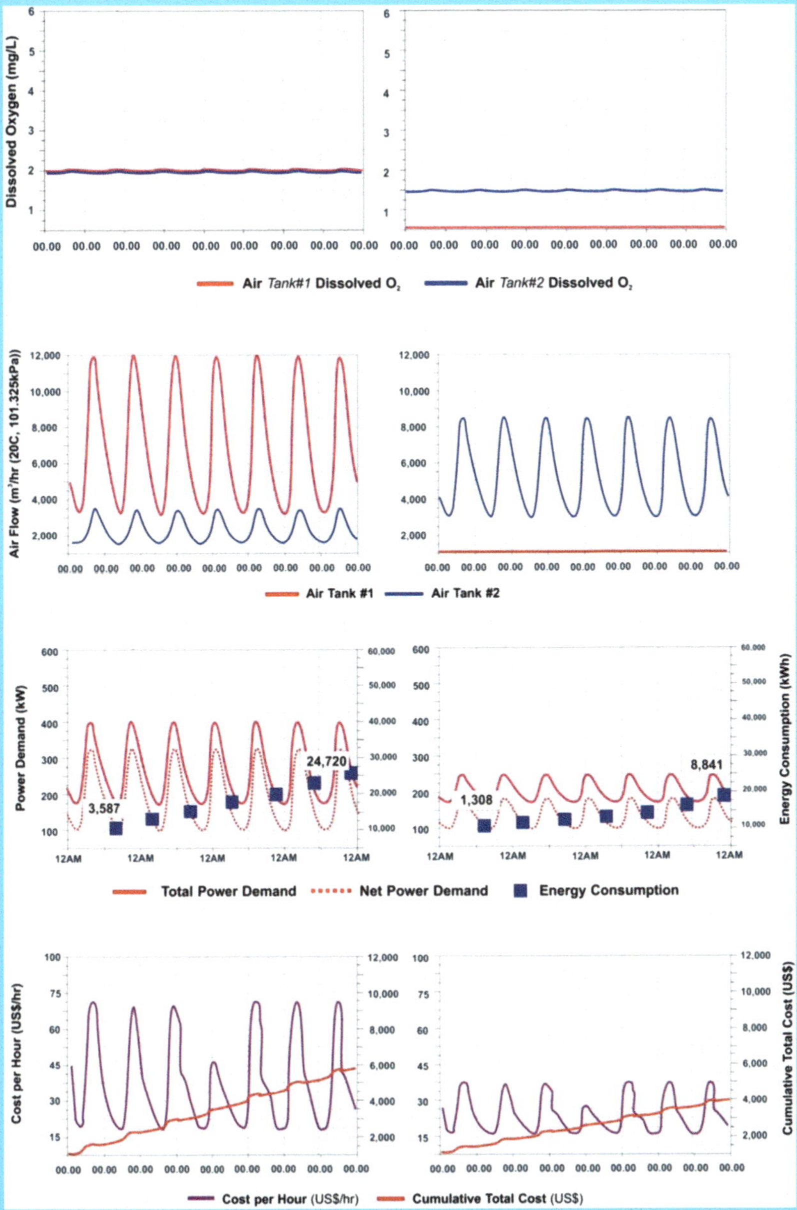

Figure 7.21 Comparison of the simulation results between the two operational strategies. On the left column is the scenario for DO = 2.0 mg/L for both tanks, while on the right is the scenario of DO = 1.5 mg/L in the first tank and zero for the second tank.

reactor from 2 mg/L to 1.5 mg/L, is presented as an example. These modifications will be compared to the base option of having both tanks aerated with a DO setpoint of 2 mg/L. In the following figures, the base option of 2 mg/L is presented on the left-hand side column, while on the right is the scenario of DO = 1.5 mg/L in the first tank and zero for the second tank.

Turning off the air in the first tank (red line), makes the airflow rate in the second tank (blue line) increase (right-hand side graphs). However, the increase is airflow rate does not reach the level observed in the baseline option. This is due to several reasons: higher α factor in the second tank and denitrification credit provided by nitrate reduction in the first tank.

The resulting effect in the net overall power demand and energy consumption at the plant is obvious, as it can be seen in the figure above. The net energy consumption (total consumption minus electricity generated from biogas) in the baseline situation over the week of simulation is 24,720 kWh vs. 8,841 kWh when operating with the air off in the first tank (right-hand side graph).

The lower energy consumption and power demand results in lower costs for the plant: US$2,100 savings per week, which represents US$110,000 savings per year, approximately. Note that a remarkable reduction in operating costs could be achieved with no capital expenditure.

The dynamic simulator can be used to determine unit costs for different processes (all considered in the analysis above). The preceding graphs show the hourly costs for the last hour of operation during the simulation. The cost in aeration is reduced to almost half from the baseline (unit costs of other unit operations remain identical, thereby reducing the effect of this intervention in the total net cost for the whole plant).

The obvious question that remains to be answered in this comparison is whether the changes introduced to the plant had an impact on effluent quality. Table 7.4 presents average effluent quality values for the baseline situation (left-hand side column) and the plant operating with the first tank un-aerated while keeping the second tank with 1.5 mg/L of DO.

As it can be seen from this table, while the ammonia-N is slightly higher in the second case (still under 1.5 mgN/L), the total nitrogen is reduced from 30.8 mgN/L in the baseline down to 18.4 mg/L, total phosphorus increases from the baseline by less than 1 mg/L, while total $CBOD_5$ is kept below 15 mg/L in both cases. In both cases, total suspended solids are above 25 mg/L. If this was an issue in terms of water quality, other options could be investigated with the simulator: lowering the SRT from 12 days to 10 days (no CapEx impact, but some potential OpEx impact due to additional biosolids disposal; increasing secondary settling surface area (both CapEx and OpEx impact); step-feeding into the second aeration tank; among several other options.

Multiple other combinations and permutations can be tested using a dynamic simulator incorporating the aeration modelling aspects discussed in this chapter. As shown in the previous examples, modelling is useful for process design as well as for decision making during the operation and

Table 7.4 Average effluent quality values for the baseline situation (left-hand side column) and plant operating with the first tank un-aerated while keeping the second tank with 1.5 mg/L of DO.

	DO Setpoint = 2 mg/l	**DO Setpoint = 1 mg/l**
Flow	13,085.00 m^3/d	13,085.00 m^3/d
Ammonia N	0.09 mgN/L	1.46 mgN/L
Nitrate N	27.15 mgN/L	11.46 mgN/L
Nitrite N	0.03 mgN/L	1.56 mgN/L
Filtered TKN	1.98 mgN/L	3.43 mgN/L
Total N	30.88 mgN/L	18.42 mgN/L
Total P	9.61 mP/L	10.52 mP/L
Total Suspended Solids	25.54 mg/L	33.95 mg/L
Total COD	50.99 mg/L	61.53 mg/L
Total Carbonaceous BOD	9.49 mg/L	13.05 mg/L
pH	7.00	7.00

maintenance of the plant. Testing these modifications to the system in a physical model (pilot plant) would take months (or years!), while simulating them in a calibrated dynamic model takes a few hours/days, thereby representing additional savings. Caution must be taken when setting up the layout of the plant in the simulator: the right physical dimensions of the unit processes must be entered, the influent must be adequately characterized, and the model parameters must be properly calibrated to match the actual operational results. Once this is accomplished, the predictions from the model under future potential operational conditions can be trusted.

7.4 FUTURE OUTLOOK OF MODELLING AERATION AND ENERGY

There are many reasons to use dynamic models of aeration and energy: increased performance, higher reliability and process stability, and a reduction in operational costs, but there are still many challenges for dynamic modelling. The oxygen transfer efficiency is a key variable in dynamic aeration and energy models, and the quantification of OTE still relies on the combination of SOTE and αF. Even if the above-mentioned mathematical model of dynamic α was developed to narrow the uncertainty on α, we have nevertheless to collect a lot of site-specific data in WRRFs, and need collect the data from different manufactures with different types of diffusers.

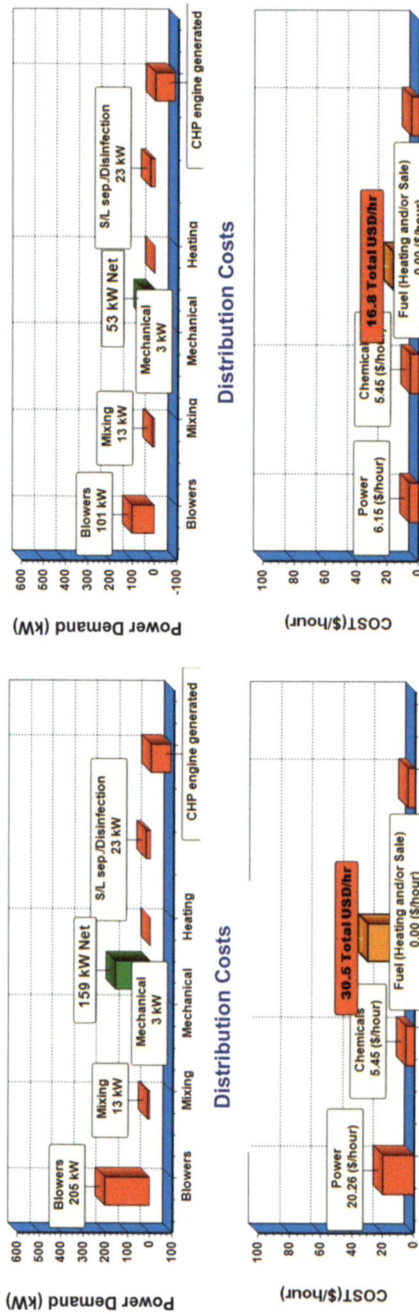

Figure 7.22 Comparison of the power demand and the operational costs between the two DO strategies. On the left column is the scenario for DO = 2.0 mg/L for both tanks, while on the right is the scenario of DO = 1.5 mg/L in the first tank and zero for the second tank.

We need pooling all WRRF oxygen-related measurements to portray a picture of the impacts of selected processes on aeration efficiency and of the most important parameters affecting aeration efficiency, not only the time-dependent variables but also the spatially dependent variables. This work seems impossible to carry out before we design a new WWTP, leading to the use of constant α factors as the most common practice in process design and modelling.

The future trend in dynamic modelling will be blending empirical data with process models in software platforms. Researchers tried to introduce reliable methods to collect the empirical data, correlate the dynamic α to the existing data sets and estimate the correlation uncertainty. Once the uncertainty is under an acceptable threshold, they integrated such relationship into a process model to evaluate broad applications. It is very difficult for engineers to collect a lot of data from different plants and diffusers before they set up this types of models, hence the reliance on extrapolation seems inevitable.

The process models (i.e., mechanistic models) are the product of abstraction of our understanding on the aeration process and so far include: bubble formation and dynamic evolution to determine the size distribution of bubbles which will affect the interfacial area between bubbles and liquid film; liquid film characteristics which mainly determine the coefficient k_L; bulk liquid characteristics integrating the nutrients and biomass influence which can also trigger a modification to k_L; and so on. And due to lack of validation, for the k_L and bubble size distribution we could find many mechanistic or process-based models, some of them derived from basic physical equations in the form of differential equations. It is very difficult to apply many of these models in practice.

We are now in a Fourth Industrial Revolution where artificial intelligence is permeating our society. We can use AI in modelling aeration and energy through deep learning and cloud computing, driven by the availability of vast amounts of data. We may use artificial neural networks and cloud computing to integrate empirical data-based models with process-based mechanistic models, thus reducing the limitations of both to a minimum.

7.5 REFERENCES

Amaral A., Schraa O., Rieger L., Gillot S., Fayolle Y., Bellandi G., Amerlinck Y., Mortier S. T. F. C., Gori R., Neves R. and Nopens I. (2016). Towards advanced aeration modelling: from blower to bubbles to bulk. *Water Science and Technology*, **75**, 507–517.

Amerlinck Y., De Keyser W., Urchegui G. and Nopens I. (2016). A realistic dynamic blower energy consumption model for wastewater applications. *Water Science and Technology*, **74**, 1561–1576.

Andrews J., Briggs R. and Jenkins S. (1974). *Instrumentation, Control, and Automation for Wastewater Treatment Systems, (Progress in Water Technology vol. 6)*. IAWPR/Pergamon Press, Oxford, UK.

Aymerich I., Rieger L., Sobhani R., Rosso D. and Corominas L. (2015). The difference between energy consumption and energy cost: modelling energy tariff structures for water resource recovery facilities. *Water Research*, **81**, 113–123.

Baquero-Rodríguez G. A., Lara-Borrero J. A., Nolasco D. and Rosso D. (2018). A critical review of the factors affecting modeling oxygen transfer by fine-pore diffusers in activated sludge. *Water Environment Research*, **90**, 431–441.

Barker P. and Dold P. L. (1997). General model for biological nutrient removal activated-sludge systems: model presentation. *Water Environment Research*, **69**, 969–984.

Capela S., Héduit A. and Roustan M . (2002). Influence of the water depth on fine bubble aeration efficiency in the presence of surfactants.

De Clercq B., Coen F., Vanderhaegen B. and Vanrolleghem P. A. (1999). Calibrating simple models for mixing and flow propagation in waste water treatment plants. *Water Science and Technology*, **39**(4), 61–69.

DHI (2014). *WEST Models Guide*. DHI, Høsholm, Denmark. http://www.mikebydhi.com/ (accessed February 2014).

Fisher M. J. and Boyle W. C. (1999). Effect of anaerobic and anoxic selectors on oxygen transfer in wastewater. *Water Environment Research*, **71**, 84–93.

Germain E., Nelles F., Drews A., Pearce P., Kraume M., Reid E., Judd S. J. and Stephenson T. (2007). Biomass effects on oxygen transfer in membrane bioreactors. *Water Research*, **41**, 1038–1044.

Gillot S. and Héduit A. (2008). Prediction of alpha factor values for fine pore aeration systems. *Water Science and Technology*, **57**, 1265–1269.

Gillot S., Capela-Marsal S., Roustan M. and Héduit A. (2005). Predicting oxygen transfer of fine bubble diffused aeration systems – model issued from dimensional analysis. *Water Research*, **39**, 1379–1387.

Gonzalez R. and Kamrani A. (2001). A survey of methodologies and techniques for data mining and intelligent data discovery. In: *Data Mining for Design and Manufacturing. (Massive Computing, vol 3)*, D. Braha (ed.), Springer, Boston, MA, USA.

Grady C. P. L., Daigger G. T. and Lim H. C. (2011). *Biological Wastewater Treatment*, C. P. Leslie Grady, Jr., Glen T. Daigger, Nancy G. Love and Carlos D. M. Filipe (eds), 3rd edn, Marcel Dekker, New York, USA.

Groves K. P., Daigger G. T., Simpkin T. J., Redmon D. T. and Ewing L. (1992). Evaluation of oxygen transfer efficiency and alpha-factor on a variety of diffused aeration systems. *Water Environment Research*, **64**, 691–698.

Henze M. (2000). *Activated Sludge Models ASM1, ASM2, ASM2d and ASM3*. IWA Publishing, London, UK.

Henze M., Grady C. P. L., Jr. and Gujer W. (1987). A general model for single-sludge wastewater treatment systems. *Water Research*, **21**, 505–515.

Leu S.-Y., Rosso D., Larson L. E. and Stenstrom M. K. (2009). Real-time aeration efficiency monitoring in the activated sludge process and methods to reduce energy consumption and operating costs. *Water Environment Research*, **81**, 2471–2481.

Libra J. A., Sahlmann C., Schuchardt A., Handschag J., Wiesmann U. and Gnirss R. (2005). Evaluation of ceramic and membrane diffusers under operating conditions with the dynamic offgas method. *Water Environment Research*, **77**, 447–454.

Metcalf & Eddy (1979) *Wastewater Engineering: Collection, Treatment, Disposal*, 2nd edn. McGraw-Hill, New York, NY, USA.

Plano S., Jiang L.-M., Nopens I., Weijers S. and Rosso D. (2011). Increasing the accuracy of energy footprint predictions for activated sludge processes using a rigorous dynamic aeration model. *Proceedings of the Water Environment Federation*, **2011**, 6500–6511.

Rosso D. and Stenstrom M. (2005a). Economic implications of fine pore diffuser aging. In: *Proceedings of the Water Environment Federation (WEFTEC2005)*, pp. 2465–2477.

Rosso D. and Stenstrom M. K. (2005b). Comparative economic analysis of the impacts of mean cell retention time and denitrification on aeration systems. *Water Research*, **39**, 3773–3780.

Rosso D. and Stenstrom M. (2006). Economic implications of fine-pore diffuser aging. *Water Environment Research*, **78**, 810–815.

Rosso D., Jiang L.-M., Pitt P., Hocking C. S., Stenstrom M. K., Murthy S., Hayden D. M., Zhong J., Coller D. H., Kim A.Y. and Xu H. (2013). Methodology for in situ column testing to improve accuracy during design and specification of aeration systems. *Journal of Environmental Engineering*, **139**, 530–537.

Sardeing R., Painmanakul P. and Hébrard G. (2006). Effect of surfactants on liquid-side mass transfer coefficients in gas–liquid systems: a first step to modeling. *Chemical Engineering Science*, **61**, 6249–6260.

Schraa O., Rieger L., Alex J., Schraa O., Rieger L. and Alex J. (2017). Development of a model for activated sludge aeration systems: linking air supply, distribution, and demand. *Water Science and Technology*, **75**, 552–560.

Snowling S. (2014). *GPS-X Technical reference*. Hydromantis Environmental Software, Hamilton, Ontario, Canada.

Steinmetz H. (1996). *Einfluss von Abwasserinhaltsstoffen, Stoffwechselprozessen und Betriebsparametern von Belebungsanlagen auf den Sauerstoffeintrag in AbwasserBelebtschlamm-Gemische*. Thesis, Kaiserslautern University, Germany (ISBN 3-925178-13-9).

Stenstrom M. K. and Gilbert R. G. (1981). Effects of alpha, beta and theta factor upon the design, specification and operation of aeration systems. *Water Research*, **15**, 643–654.

Wagner M. R. and Pöpel H. J. (1998). Oxygen transfer and aeration efficiency - Influence of diffuser submergence, diffuser density, and blower type. *Water Science and Technology*, **38**(3), 1–6.

Zlokarnik M. (1980). Koaleszenzphänomene im System gasförmig/flüssig und deren Einfluß auf den O2-Eintragbei der biologischen Abwasserreinigung (Coalescence phenomena in gas/liquid systems and its impact on oxygen transfer in biological wastewater treatment). *Korrespondenz Abwasser*, **27**, 728–734.

Chapter 8

Sustainable aeration practice: Design and diagnostics

Diego Rosso[1], Michael K. Stenstrom[2], José Jimenez[3] and Manel Garrido-Baserba[1]

[1]*University of California, Irvine*
[2]*University of California, Los Angeles*
[3]*Brown and Caldwell*

8.1 INTRODUCTION

Diagnostics, the art and practice of studying problems, is a fascinating field of endeavour. In this chapter we present a brief summary of our experience identifying what does not work and how to use this knowledge to improve aeration design and retrofit. Clearly, more experience and wisdom are spread throughout the industry. Hence, we do not pretend that our presentation here is the systematic and exhaustive compilation of knowledge on aeration diagnostics.

8.2 EQUIPMENT SPECIFICATION: *IN SITU* DATA COLLECTION

Earlier in the book we described how to collect data *in situ* for site-specific alpha quantification. Usually, this is much smaller than 1% of the design budget, but it confines substantially the uncertainty during the design procedure. More, *in situ* alpha measurements can be conducted in parallel to the initial phase of the design procedure, usually during the contractual phase (Rosso *et al.*, 2011).

Editor: Diego Rosso
doi: 10.2166/9781780407845_0255

We recognize that performing field work, often in challenging weather conditions, is unpleasant and frequently dangerous. It is of no surprise that most would avoid it, whenever possible. However, some have dedicated their career to this specialty and decided to maintain their role as independent consultant to provide the key service of third-party verification.

Furthermore, in litigious societies such as the United States, the benefit of third-party verification is often part of the due diligence that a licensed engineer should respect in decision-making to avoid assuming the liability for unverified claims by a manufacturer or a colleague.

Below is are some examples of *in situ* studies to measure the gas transfer coefficient in clean and process water at a full-scale treatment plant. The sites illustrated in Figure 8.1 span from small columns to large pilot units (flow ~4000 $m^3\ d^{-1}$ or ~1 MGD). The first step is the establishment of a baseline through clean water tests in the same tank where the off-gas tests for alpha would be later performed. Clean water tests may already be available from a shop test performed by the aeration system manufacturer, however the limits of the *in situ* geometry may affect the k_La measurements, therefore requiring site-specific clean water tests.

Figure 8.1 Examples of *in situ* reactors for measurements of site-specific values of alpha as a function of process variables and parameters. On the bottom left is a pilot plant and on the bottom right is a sidestream column, both undergoing off-gas tests.

Figure 8.2 Examples of modular off-gas hoods constructed from plastic elements. On the left is the hood before deployment in a pilot unit, on the right is another hood used during a full-scale test.

After the clean water tests, off-gas tests should be performed. These tests should span the air flow per diffuser correspondent to the expected process conditions. The collection of the off-gas can occur by covering the entire surface of the testing reactor or by constructing a hood. The hood, due to the small geometry of the reactor, does not need to have the same size of typical off-gas hoods for full-scale tests, but rather should be constructed to sample a sufficient fraction of the aerated area but to guarantee manoeuvrability by the testing team.

A simple way to construct off-gas hood is a modular construction using plastic elements. Figure 8.2 illustrates some examples of modular hoods. These can be expanded as needed to perform full-scale tests as well.

8.3 EQUIPMENT PERFORMANCE MONITORING AND MAINTENANCE

Fouling is inevitable and inexorable, and the extent of it is what should be quantified. In principle, there is no aeration system that is better than all the others, but it is rather a matter of applicability of a certain technology to a particular project that makes it the best solution. For example, when efficiency is the driver for decision making (such as, for example, in municipal facilities), fine-pore diffusers tend to be the selection of choice. However, this is would be an appropriate choice only insofar as the oxygen requirements are relatively low in absolute value. In the case of highly loaded processes (e.g., industrial facilities), on the other hand, the choice is driven by the need to deliver the required oxygen, often at the expense of efficiency. *The more efficient is the diffuser, the more the effect of fouling would weigh on its performance.* An entertaining analogy that can be used to remember this recommendation is the following: for one who cannot commit to change the tyres of a car with the required frequency, a bulldozer with steel tracks would be a viable vehicle (not very efficient but without any need for tyre replacement). Hence, the need to commit to understanding and monitoring fouling becomes pressing for the facilities adopting higher-efficiency technologies.

Let us take another look at the evolution of fouling over the life of the diffuser technology. Figure 8.3 illustrates how fouling affects the oxygen transfer efficiency. The initial value, for new diffusers, is $\alpha SOTE_0$, because the fouling factor F is by definition 1.0 at the initial time, and the airflow at which the diffusers operate is AFR_0. Note that this whole example is based on nominal airflows, their correspondent efficiencies and pressure drops (see below). In reality, diffusers operate over a range of airflows, and the considerations here should be applied to the entire range of operation. Furthermore, each grid along the treatment train will experience different extents of fouling, with more severe effects on the initial grid and much less effect on the final grid. For sake of this example, thus, we should consider the AFR here as the tank average airflow and $\alpha SOTE$ as the airflow weighted average $<\alpha SOTE>$. At this point in time, the oxygen requirements, which are proportional to the product of airflow and efficiency are represented by the light-blue rectangle.

After some time in operation, the diffusers experience fouling, which corresponds to a sag of the entire operating curve $\alpha SOTE$ vs. AFR (in this case the curve becomes $\alpha FSOTE$ vs. AFR). The shape of the curve in Figure 8.3 is maintained, but experience shows that there is a more dramatic decrease at lower airflow rates and a more dramatic decrease for the first grid of a tank with grids in series. All curves can be quantified by off-gas analysis.

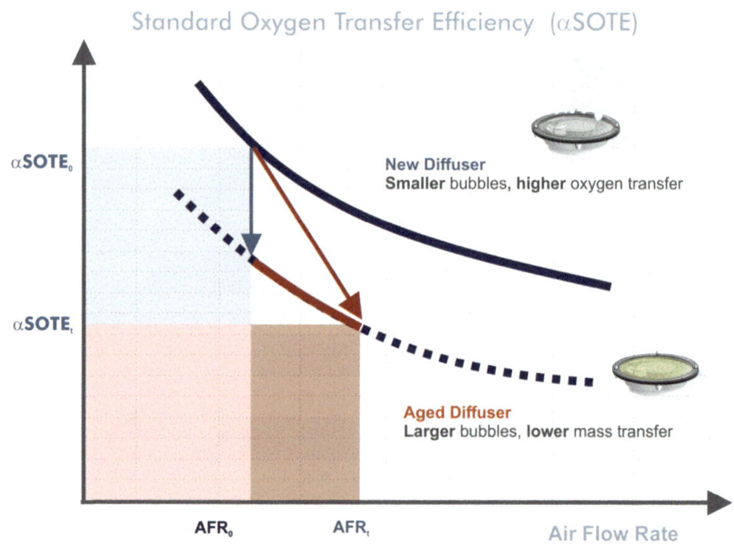

Figure 8.3 Evolution of the oxygen transfer efficiency in process water $\alpha SOTE$ (and $\alpha FSOTE$) and of the fouling factor F for fine-pore diffusers over time in operation.

After the given time in operation, the efficiency has sagged to a lower value but the aeration system controller does not see that and only records a decrease in DO for the AFR_o. Therefore, an increase in airflow is necessary to maintain the DO and meet the oxygen requirements that are still the driver for the process. The oxygen requirements at time *t* are represented by the red rectangle, whose area is same as that of the light-blue rectangle. Hence, the compounding effect of the decrease in efficiency at a given airflow, and the increase in airflow necessary to front the decrease in efficiency (hence, the decrease in oxygen transfer rate), amounts to a much increased airflow (AFR_t in Figure 8.3).

Meanwhile, the ensuing fouling caused an increase in pressure drop DWP (Figure 8.4). However, the pressure drop increase that the blowers need to face is that corresponding to the AFR_t, needed to deliver the required oxygen for treatment. Since the blower power requirements are proportional to AFR and to $DWP^{0.283}$, the compounding effect of fouling results in a more than linear escalation in power demand.

A very good practice, coming at a very small cost to the plant operators and engineers, is to track the amount of air discharged per unit load resolved by the process. An increase with months in operation of the air volume required per unit load treated (example in Figure 8.5) would be associated with the erosion of oxygen transfer efficiency, quantifiable by a decrease in the fouling factor *F* and by an escalation of the power drawn by the blower house (example in Figure 8.6).

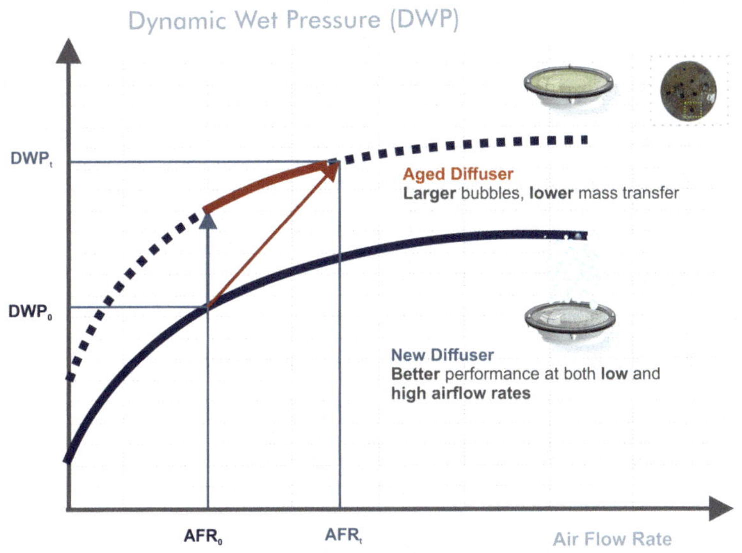

Figure 8.4 Evolution of the diffuser pressure drop DWP and of the pressure factor Ψ for fine-pore diffusers over time in operation.

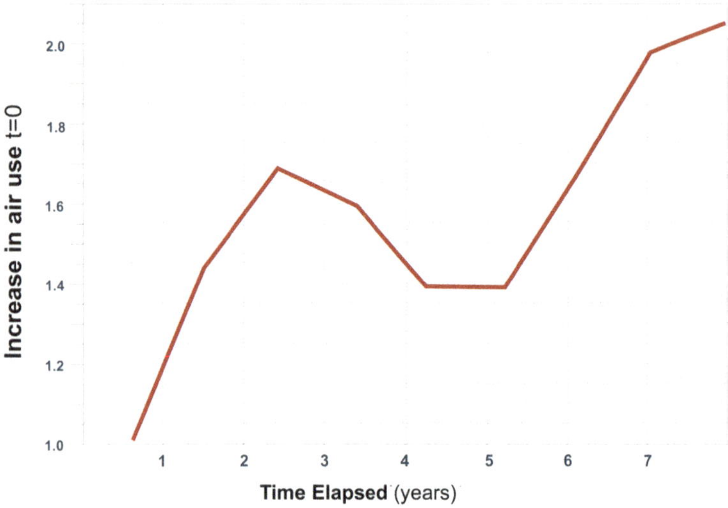

Figure 8.5 Relative increase in air use per unit load treated by a full-scale treatment facility over an 8 year period. This plot was generated from process data logs.

Another relatively easy practice is to track the discharge pressure from the blower house. However, owing to the high proportion that the hydrostatic head has on the overall discharge pressure, this practice requires good-quality pressure

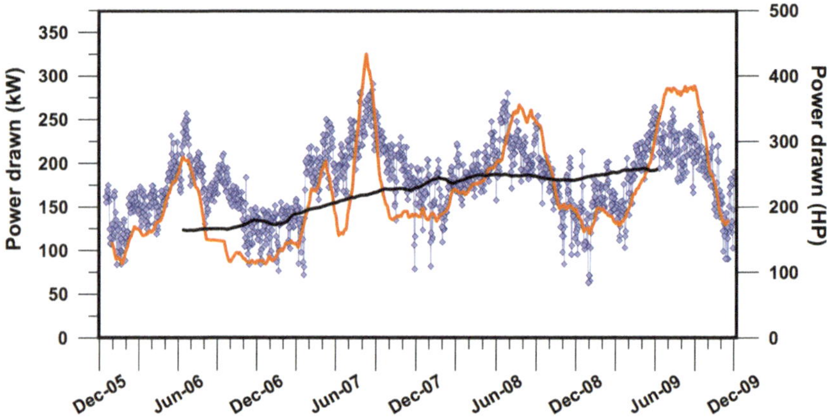

Figure 8.6 Power drawn by the blower house of a full-scale wastewater treatment plant over a period of five years. Note the yearly seasonal cycle, shown by the daily values (blue points) and the 30-day running average (orange line). The 1-year running average (black line) shows the long-term increase in power demand, almost doubling in the period covered by this plot. The calculations to produce this plot were based on process data logs.

gauges and can be applied to diffusers whose pressure drop will increase by an absolute value that can be recorded within the discrimination of the gauge. Therefore, it would be very challenging if not impossible to notice the pressure drop increase of ceramic diffusers.

Diffuser cleaning can be performed by:

- mechanical cleaning: tank-top hosing; pressure washing; brushing; pneumatically (flexing/reverse-flexing membranes)
- chemical cleaning using: acids (liquid or gas), bases, detergents

Current knowledge suggests that cleaning once every 24 months is a good practice to curb the decline in efficiency of all diffusers (Rosso & Stenstrom, 2006), however cleaning effectiveness and frequency is site-specific and should be evaluated on a case by case basis. When using chemicals, care must be applied to prevent damage of the air distribution piping and fittings. Tank-top hosing and pressure washing should be performed with some water above the diffusers, to prevent mechanical damage or destruction of diffuser membranes.

Regarding the effectiveness of pneumatic cleaning, a study on diffusers that were flexed and reverse-flexed (i.e., the pressure in the air feed was relieved, causing membrane collapse onto the diffuser frame) was performed (Figure 8.7). In that

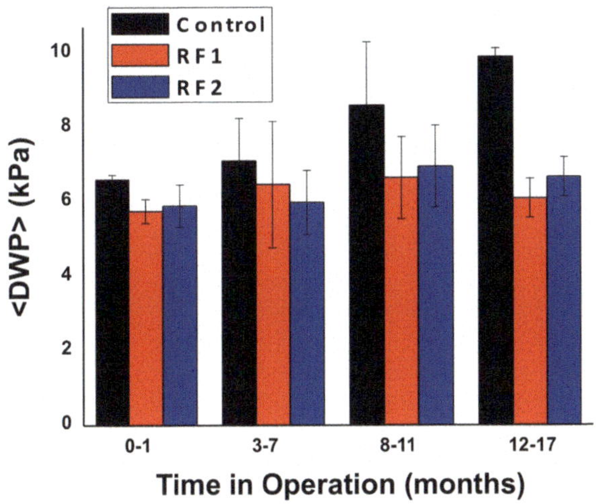

Figure 8.7 Airflow-weighted average of the discharge pressure <DWP> of polyurethane membrane panels over time in operation. The plot shows both the pressure drop of a diffuser not experiencing the prophylaxis of flexing and reverse flexing its membrane (labelled Control), and of diffusers undergoing such procedure daily (labelled RF1 and RF2). The benefit of flexing the membranes is here clear, despite the lack of effectiveness in curbing the decline in oxygen transfer efficiency for these particular diffusers (data from Odize *et al.*, 2017).

case, pneumatic cleaning proved effective in curbing the increase in DWP, however the decline in αFSOTE showed no statistically significant difference between diffusers that were pneumatically cleaned and those that were not. Nevertheless, this practice is automated and virtually effortless (assuming that the blower discharge pressure can actually meet the requirements for effective flexing) and even if only the escalation of DWP were curbed, the operation of the blower battery may avert additional units coming on line at peak power demand charge periods.

We present here three case studies performed on municipal wastewater treatment plants to show the implementation of possible energy-conservation strategies. A summary of the data regarding the treatment plants is set out in Table 8.1.

Table 8.1 Background of tested wastewater treatment plants.

	Plant 1	Plant 2	Plant 3
Average volumetric flow (m^3/hour)	10,000	3150	12,500
BOD_5 in primary effluent (mg/L)	162	132	146
Ammonia in primary effluent (mg-N/L)	28	25	28
Tanks in operation	18	6	10
Treatment layout	NDN[a]	NDN	CAS[b]
Population served	800,000	220,000	880,000

[a]NDN = nitrification/denitrification process.
[b]CAS = conventional activated sludge process.

Plant 1 – This treatment plant serves 800,000 people and treats approximately 240,000 $m^3\ d^{-1}$. This plant uses ceramic disc diffusers for aeration, installed as retrofit from a previous system. During that retrofit, the treatment operations were upgraded from conventional activated sludge process (CAS) to nitrification/ denitrification (NDN). Off-gas tests were applied continuously in two processes to evaluate diffuser fouling.

Plant 2 – The operation and testing scenario of this plant is similar to Plant 1, with ceramic disc diffusers for aeration and treatment converted from conventional to NDN. The capacity of this plant is smaller than the other two: the population served is approximately 220,000 and the average volumetric flow rate of this plant is 76,000 $m^3\ d^{-1}$.

Plant 3 – Fine-pore diffusers used in this treatment plant are EPDM membranes, and this plant has not yet been upgraded to NDN process due to overloading. Similar to Plant 1, the capacity of this plant is 12,500

$m^3\ d^{-1}$ and the population served is 880,000. The diffusers in more than three-quarters of the tanks have not been cleaned in eight years.

Blower power was quantified using the adiabatic formula of blowers (Metcalf & Eddy, 2014). αSOTE for all processes decreases with time in operation. However, the rate of fouling/scaling varies with diffuser types and plant operation. The fouling is extremely fast when MCRT is short, due to the more aggressive type of fouling associated with higher-rate processes (as reported by Garrido-Baserba *et al.*, 2017).

Figure 8.8 shows the aeration power costs in these three plants. The diffusers were grouped and colored by fouling status. The bar on top of each column shows the standard deviation of test results. Significant improvements on energy usage by diffuser cleaning were observed, which provided an energy saving to approximately 4.5 kWh y^{-1} per capital, or 9900 kWh d^{-1} in Plant 1.

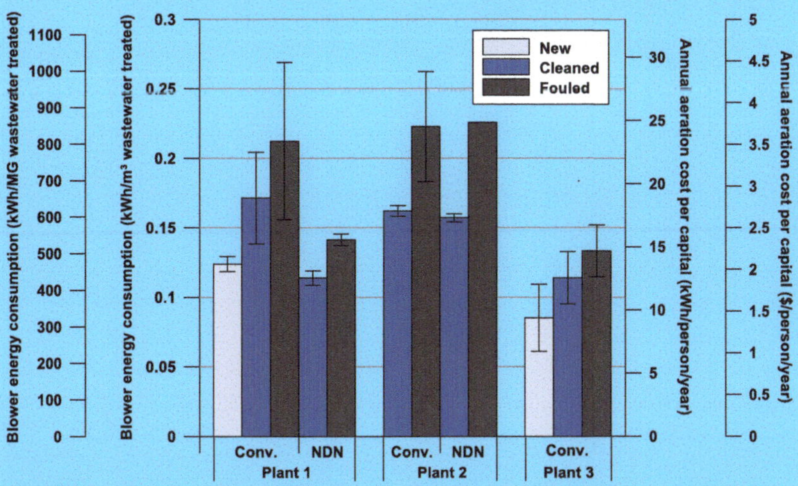

Figure 8.8 Comparative energy consumption of fine-pores diffusers in different treatment plants. All aeration costs were normalized per unit load and population served per year.

In addition to cleaning diffusers, one important prophylaxis is to ensure good grit removal. Why are talking about grit removal in a book on aeration? Because without good grit removal upstream of the aeration tank, the accumulation of grit in the tank could cripple the aeration diffusers, especially in grids with low air flux (Figure 8.9).

Figure 8.9 Example of the deleterious effects of grit accumulation on the aeration tank floor, with corresponding obstruction and damage to the diffusers.

8.4 ROLES AND TERM DEFINITIONS: WHO IS DOING WHAT

Several parties are included in the design procedure of aeration systems, as illustrated in Figure 8.10. The main role of the manufacturer is to supply the aeration system and to verify its performance using a factory or field test (under the witnessing eyes of an independent third-party expert) in a repeatable and transparent fashion. This is why standards for testing aeration systems in clean and process water are crucial. In an analogous fashion, blower performance and efficiency should be assessed using the separation of roles illustrated here.

The third party witness must bear no conflict of interest with the manufacturer, of course. Also, the consultant who designs the aeration system should request the third-party verification whenever in-house expertise to critically evaluate the equipment warranty is not available (the devil is in the details, as the saying goes). Since the manufacturer best knows the aeration technology provided, they need to guarantee the clean water performance, which is a measure of the equipment's performance and efficiency in a standard condition that is independent of the wastewater. The ratio between oxygen transfer rate in process water and clean water (the mysterious *alpha factor*) is a function of influent characteristics and operating conditions more than of equipment. Given that understanding the wastewater characteristics and operating conditions is the responsibility of the consulting engineers in charge of process design, it would be their role to select the ranges of alpha factors. The consultant who uses an alpha

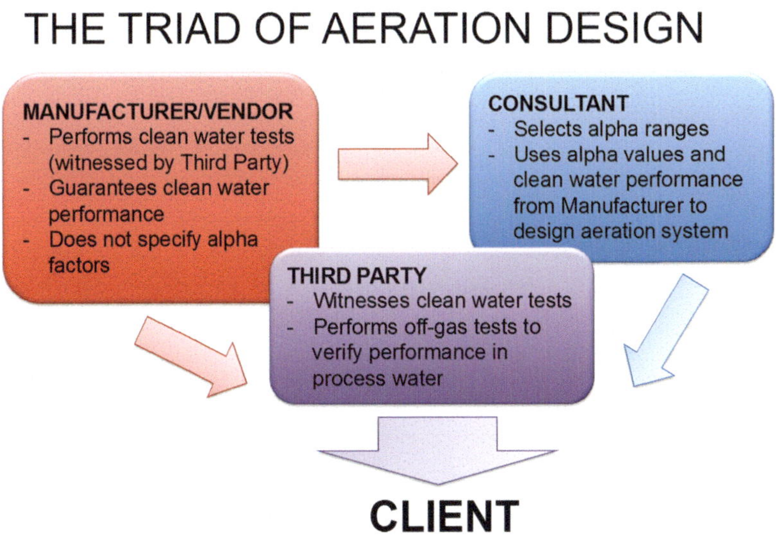

Figure 8.10 Roles and responsibilities during an ideal aeration system design. The arrows broadly describe the flow of information during the project.

factor suggestion from the manufacturer without critically addressing its validity is exposing the design team to the potential failure of the system (and probable litigation with the client). In fact, the consulting engineers are hired specifically to help the client decipher the quantitative implications of design assumptions: otherwise, the client could likely carry out the design on their own. If all the roles are separate and the aeration performance is tested according to standard protocols (e.g., the ASCE or European Standards), the design procedure should yield a fully functional aeration system whose performance and efficiency can be verified independently and be within the targeted expectations. The role of the third-party expert is to verify the manufacturer data and to verify the consultant assumptions on alpha.

It is of crucial importance to clearly define the efficiency and performance parameters of aeration systems during their analysis or specification. To compare different technologies, as well as to monitor the same aeration system over extended time in operation, a level-field comparison is necessary. The most basic parameter to define *performance* is the OTR. This is the design target: should the amount of oxygen transferred to the water per unit time be lower than the target OTR for extended periods of time each day (or especially indefinitely), the biological reactor would experience oxygen deficiency and its performance would be compromised and effluent discharge violations would be likely.

The main parameter used to quantify the *efficiency* of the oxygen transfer process is the OTE. In order to avoid bias due to site-specific environmental and process conditions, the corrections to standard conditions should be carefully calculated. Methods for translating standard to process conditions for all situations (i.e., barometric pressure, temperature, etc.) are shown in detail in the standard protocols cited here.

When defining the overall aeration efficiency for surface aeration, a quantification of the power invested in the aeration process is needed. This is a case when the engineer specifying the aeration system should be familiar with electromechanical technology and specification, or a competent engineer on electrical technology should be asked to join the team.

8.5 AERATION DIAGNOSTICS: WALKING AROUND THE PLANT

A very important component of aeration diagnostics is walking the plant looking for symptoms of unhealthy conditions. An example of simple fault detection is the observance of surface boils, similar to small geysers, which indicate damaged diffusers or leaking air headers (Figure 8.11).

One may be sufficiently lucky to encounter a dewatered tank at a treatment facility. Certain details, visible only when the tank is dewatered partially or in full, allow for an inspection of the aeration system elements that the full operating tank would not. For example, the selective replacement of one cracked diffuser header (Figure 8.12) produces differential air discharge due to the lower pressure drop associated with the new diffusers.

Another issue that can be identified with visual observation of partially dewatered tanks is the uniform discharge of air throughout a diffuser grid. In Figure 8.13 we show a patchy air distribution from a group of diffusers in a new installation, likely due to poor header levelling during on site assembly.

Figure 8.11 Surface boil indicating on the surface that at the bottom of the tank a diffuser is either damaged or an air header is leaking.

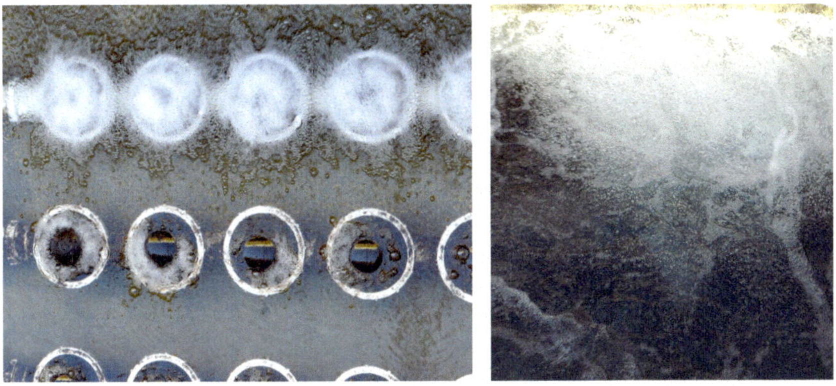

Figure 8.12 Differential air discharge from a new diffuser header installed to replace a damaged header. Note the increased air discharge from the new diffusers, due to their lower pressure drop. The image on the left shows the diffusers and the image on the right is the surface of the tank.

Finally, one very important element, which is rather negligible in cost but important in its functionality, is a properly installed purge to drain continuously the moisture that may accumulate and the sludge that may enter the air headers. This purge can be opened periodically to drain the collected liquid or can be always left open. Ideally, this purge would be a tube ~10–20 mm in diameter connecting the top of the tank above water with the bottom of the aeration header

Figure 8.13 Patchy air distribution from a group of diffusers in a new installation, likely due to poor header levelling during assembly.

(i.e., the low point where liquid would inevitably accumulate). A good design procedure should always include a purge.

8.6 PROCESS STABILITY BENEFITS OF GOOD AERATION PRACTICES

A good design procedure, especially when retrofitting existing systems to configurations with internal recirculation, is to avoid proximity of diffusers with the intake of the internal recirculation pump. This is to avoid the entrainment of bubbles or the intake of mixed liquor with elevated DO in the pump line, which would return it to anoxic tanks where it would cause cripple denitrification and provoke filamentous events.

Also, the overzealous aeration practice of setting excessive DO levels may lead to excessive shearing of the sludge by the bubbles, which could trigger sporadic if not systematic solid separation issues. Much research is still needed to elucidate the role of aeration on the well being of the biokinetic metabolism as it pertains solids separation in secondary clarifiers. As remarked by Wahlberg (2013) in his WEFTEC lecture, the key for a healthy secondary clarifier is a well functioning biological process.

8.7 MECHANICALLY SIMPLE AERATION SYSTEMS

Lagoons or stabilization ponds are often inexpensive methods for wastewater treatment with low energy intensity. However, their land intensity is elevated and therefore their suitability as a sustainable alternative relates mainly to small communities and/or rural areas. For facultative water resource recovery lagoons, the driver for oxygen transfer and for process energy consumption is mixing. Figure 8.14 illustrates a schematic of an aerated lagoon.

In the recent past, the price of photovoltaic batteries has decreased sharply, facilitating the replacement of existing mechanical mixers with solar-powered mixers. Jiang *et al.* (2018) studied a case of a lagoon undergoing parallel evaluation of existing grid-powered mechanical mixers and a solar-powered mixer, showing the adequacy of the solar mixer (Fig. 8.15). In that case (population served <1000), both energy usage and power cost exhibited a significant reduction after the solar-powered mixer was installed: energy usage decreasing 70% and power cost decreasing by 47% in the quarter studied. For locations where regulatory incentives are present to curb carbon emissions, the transition of mixing to off-grid units would signify a reduction in indirect carbon emission from electricity importation. In this case, the decrease by 64% in carbon emission associated with electricity importation was nevertheless a small change in the overall carbon emission (from 3.8% to 1.5%), due to the elevated direct carbon emission of this process.

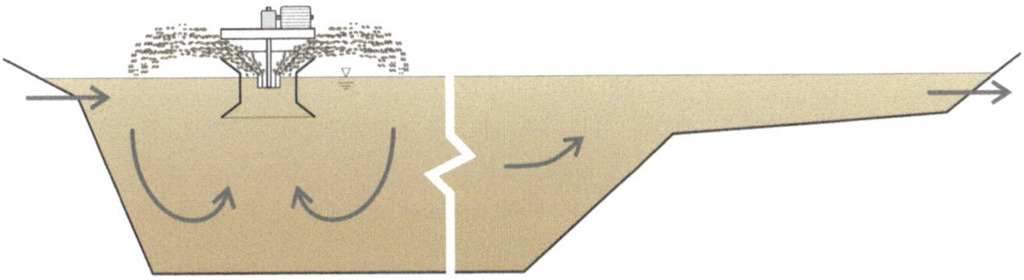

Figure 8.14 Schematic of a section of an aerated lagoon equipped with a surface aerator.

Aeration for aerated lagoons is usually performed by mechanical aeration, although diffused aeration is possible. In the case of diffused aeration (Figure 8.16), primary separation is a key element to prevent the destructive accumulation of fibrous material and rags onto the diffuser assemblies. For mechanical aeration, high-speed surface aerators are usually employed, typically operating on the low end of the SAE ranges. The additional issue of heat loss in winter, exacerbated by surface aeration systems, does contribute to the interest in retrofits to diffused aeration systems. Diffused aeration systems in lagoons tend to employ floating systems rather than diffuser grids anchored to the floor. This facilitates the retrieval of single diffuser assemblies with the service barge and adapts well the diffuser system to lagoons of uneven depth by levelling the diffusers to the surface. This strategy maintains even the discharge of bubbles

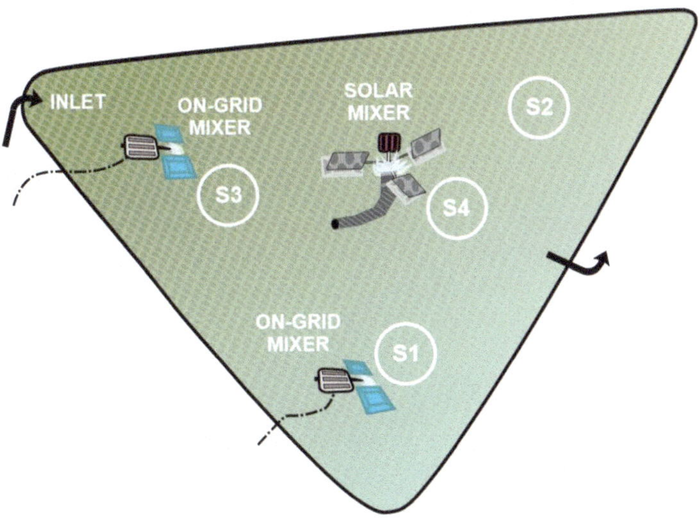

Figure 8.15 Schematic of a facultative lagoon for a small community, where a side-by-side evaluation of on-grid and solar-powered mixers was performed by Jiang *et al.* (2018).

Figure 8.16 Example of aerated lagoons in operation at a large facility. On the top is a view of the lagoon from the effluent side; on the left shore is the air manifold with all the floating headers where many diffuser assemblies are attached and floating underwater. On the bottom left is an image of the blower house. On the bottom centre is a photograph of a damaged assembly being retrieved on the service barge; note the large quantity of rags being removed from the surface of the assembly, visible on the left and top of the image; one of the diffuser membranes was removed from this assembly for off-site fouling studies. On the bottom right is a detail of said diffuser membrane surface.

across distant assemblies, since the diffuser submergence for all units is pegged to the water surface.

There are trade-offs between the maintenance requirements of the two alternative aeration methods. If surface aerators are used, they must be maintained using boats with personnel trained to work offshore, which may be prohibitive in winter. If fine-pore diffusers are used, the blower maintenance is always performed primarily onshore, but diffuser maintenance would require offshore methods such as a barge.

8.8 THE ELEPHANT IN THE ROOM: HPO PROCESSES

The High Purity Oxygen Activated Sludge Process (HPOAS; Figure 8.17) involves the use of pure oxygen in covered stage reactors, instead of air diffusers or mechanical aerators in uncovered basins. The process was developed to front the need for plant expansion in capacity when limitations were imposed on the physical footprint of the facility. Hence, most of these installations are in densely populated areas and tend to be large processes. In the most common layout, 3–5

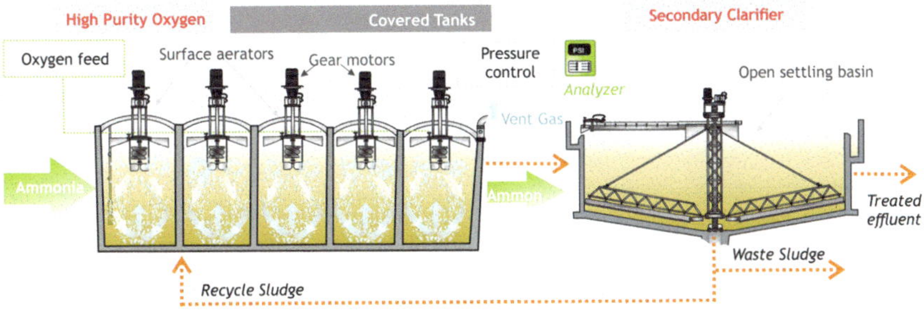

Figure 8.17 Typical high purity oxygen activated sludge process. The series of reactors is covered while the secondary clarifiers are open to atmosphere.

reactors in series are used to ensure a high utilization of the applied oxygen. The process was introduced and developed by the Union Carbide Corporation. For a contemporary history of the technology and its development program see the book by the process inventor (McWhirter & Vahldieck, 1978).

The adoption of this technology reached its maturity one generation ago (in the mid 1980s), as highlighted in Figure 8.18. The age of these facilities, combined with the increased energy costs and interest in nutrient removal/recovery, poses the question of what to do as we approach the end of their useful life. A large portion, if not the majority, of the wastewater flow in the United States is received by a small number of HPOAS facilities. More importantly, approximately

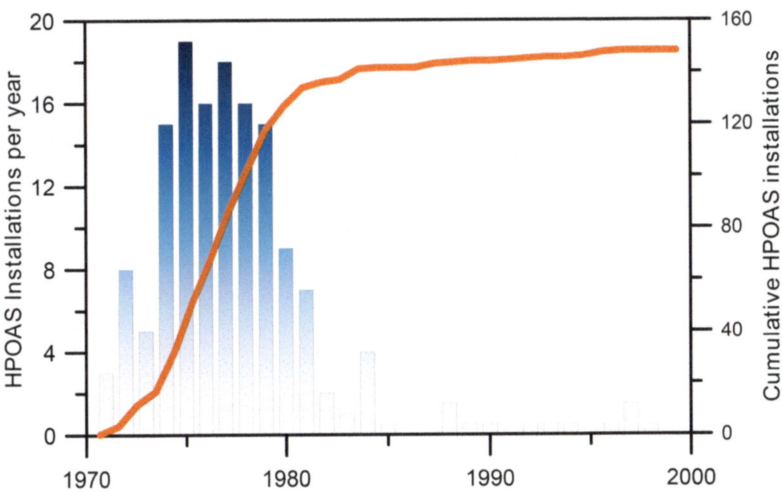

Figure 8.18 S-curve for North American High Purity Oxygen Activated Sludge plants (data from Parker, 2011). In red is the cumulative curve and in grey are the installations per year.

one-third of the United States wastewater flow is handled by the 47 plants larger than 100 MGD (Metcalf & Eddy, 2014), many of which are located in coastal areas (e.g.: New York, Boston, Philadelphia, Los Angeles, San Francisco, Miami, Seattle, etc.). In fact, given the small number of HPOAS plants and their relatively large size, the analysis of a small number of plants can impact a large number of people served on a continental scale.

HPOAS processes are almost exclusively performing BOD_5 removal, which is a challenge when nutrient removal requirements are faced by the utility. Moreover, as water reuse becomes an integrated part of the water supply portfolio in various areas, changes to the HPOAS plants appear on the horizon and the future of these HPOAS plants needs to be rethought to address the following limitations (Fig. 8.19):

- *difficulty or inability to nitrify*, and when attempted, prohibitive foaming issues have impaired the overall performance and operation of the HPO plant;
- *challenges in solids separation*, due to poor mixed liquor settling, elevated wasting rate (i.e., low MLSS) and high shear in the reactors;
- *elevated energy intensity* and escalating operating costs for oxygen production;
- effluent not *suitable for water reuse*.

A well controlled process achieves 70–80% oxygen utilization by the bacteria, which equates to 20–30% of the oxygen generated being wasted. In fact, the

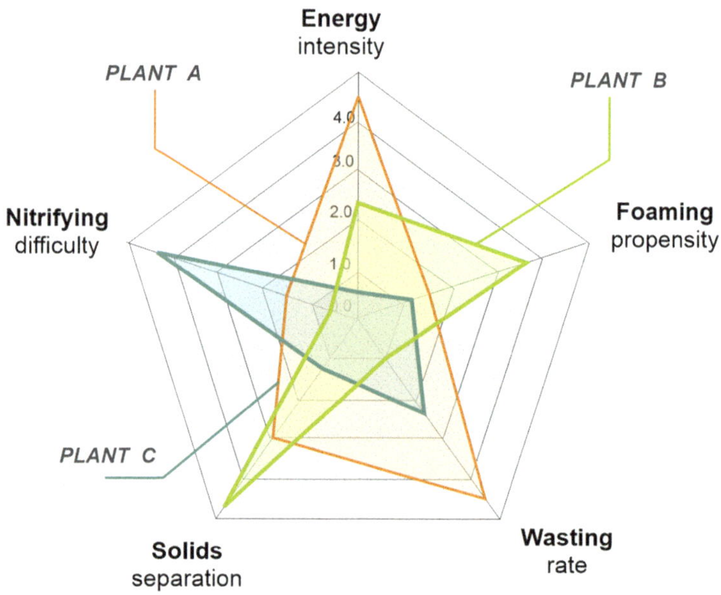

Figure 8.19 Illustration of some challenges faced by High Purity Oxygen Activated Sludge processes for three sample plants.

original estimates by Union Carbide of optimal performance reaching ~90% oxygen utilization were achieved in rare cases. However, the engineers at the San Francisco Oceanside Water Pollution Control Plant (OSP) devised a more efficient control strategy capable of achieving up to ~100% utilization of the oxygen supplied to the process (Miot *et al.*, 2016).

As many of these facilities are reaching the end of their useful life, the open question of their future is becoming ever more important. However, each case is treated on a site-specific basis drawing from this suite of these technological options:

- decommission HPOAS facility
- decommission oxygen column and purchase oxygen from *ex situ* supplier
- convert partially or completely to atmospheric process
- incorporate of state-of-the-art intensification technologies (i.e., ballasted processes, biofilm-based processes)
- nitrify and denitrify in a modified HPO process
- add a NDN process downstream from HPO.

At the time this book was written there were no clear guidelines on how to approach the issue systematically.

8.9 REFERENCES

Garrido-Baserba M., Sobhani R., Asvapathanagul P., McCarthy G. W., Olson B. H., Odize V., Al-Omari A., Murthy S., Nifong A., Godwin J., Bott C. B., Stenstrom M. K., Shaw A. R. and Rosso D. (2017). Modelling the link amongst fine-pore diffuser fouling, oxygen transfer efficiency, and aeration energy intensity. *Water Research*, **111**, 127–139.

Jiang Y., Bebee B., Mendoza A., Robinson A. K., Zhang X. and Rosso D. (2018). Energy footprint and carbon emission reduction using off-the-grid solar-powered mixing for lagoon treatment. *Journal of Environmental Management*, **205**, 125–133.

McWhirter J. R. and Vahldieck N. P. (1978). Oxygenation systems mass transfer design considerations. In: *The Use of High-purity Oxygen in the Activated Sludge Process*, J. R. McWhirter, (ed.), vol. 1, CRC Press, West Palm Beach, FL, USA, pp. 235–260.

Metcalf & Eddy (2014). *Wastewater Engineering: Treatment and Resource Recovery*, 5th edn. McGraw-Hill Education, New York, NY, USA.

Miot A., Levitin Y. and Stenstrom M. K. (2016). Novel control method to achieve higher oxygen utilization in high purity oxygen activated sludge plants. In: *Proceedings of the Water Environment Federation (WEFTEC 2016)*, pp. 3623–3631.

Odize V. O., Novak J., Rosso D., De Clippeleir H., Al-Omari A., Garrido M., Smeraldi J. and Murthy S. (2017). Reverse flexing as a physical/mechanical treatment to mitigate fouling of fine bubble diffusers. *Water Science & Technology*, **76**, 1595–1602.

Parker D. (2011). Introduction of new process technology into the wastewater treatment sector. *Water Environment Research*, **83**, 483–497.

Rosso D. and Stenstrom M. (2006). Economic implications of fine-pore diffuser aging. *Water Environment Research*, **78**, 810–815.

Rosso D., Lothman S. E., Jeung M. K., Pitt P., Gellner W. J., Stone A. L. and Howard D. (2011). Oxygen transfer and uptake, nutrient removal, and energy footprint of parallel fullscale IFAS and activated sludge processes. *Water Research*, **45**, 5987–5996.

Wahlberg E. J. (2013). What really makes activated sludge clarifiers work? In: *Proceedings of the Water Environment Federation (WEFTEC 2013)*, pp. 3987–3995.

Epilogue

Diego Rosso
University of California, Irvine

This was my first time producing an entire book. Now I look at other books very differently. Often I noticed how other books missed references I deemed important (some were my papers, some were others). After having read these chapters so many times, I honestly cannot see what is missing and what is there. If we missed a citation or some important point, let us know.

As you can see, we made an attempt at avoiding the repetition of material available elsewhere, and instead tried to critically assemble the information and knowledge from a multitude of research papers and from our experience. In fact, you do not see a detailed aeration design procedure here: we rather focus on providing feedback on the design elements that require critical attention.

The compilation of this book took years and was the effort of many experienced colleagues, who in the process learnt much from each other. As you may notice, we tried to cover all dimensions of aeration without compiling an encyclopaedia. We tried to entertain you with stories from our experience – mind there are many more where those came from: just ask the plant operators and you will find an immense source of knowledge.

I am indebted to all the co-authors who gave so much of their free time to contribute to this book. Finally, we should thank both our publishers (Maggie Smith and Mark Hammond) for their persistence in seeing our work through.

This is it.

Editor: Diego Rosso
doi: 10.2166/9781780407845_0275

Index

IWA Publishing's authorised EU representative for General Product Safety Regulations is Diane D'Arras, 15 rue Duret, 75116 Paris, France, e-mail: safety@iwap.co.uk.

Printed and bound by CPI Group (UK) Ltd, Croydon, CR0 4YY
12/05/2026
02108873-0001